CLIMATE AND ENERGY LIES

EXPENSIVE, DANGEROUS, AND DESTRUCTIVE

FRANK LASÈE

Copyright © Frank Lasèe 2024
Climate and Energy Lies
Expensive, Dangerous, and Destructive

Published by Pierucci Publishing, P.O. Box 2074, Carbondale, Colorado 81623, USA
www.pieruccipublishing.com

Cover design by Stephanie Pierucci
Edited by Stephanie Pierucci

ISBN
Ebook: 978-1-962578-36-3
Hardcover: 978-1-962578-46-2
Paperback: 978-1-962578-49-3

Library of Congress Control Number: 2024911854

All rights reserved. Except as permitted under U.S. Copyright Act of 1976, no part of this publication may be reproduced, distributed, or transmitted in any form or by any means, or stored in a database or retrieval system without the prior written permission of the copyright owner. The scanning, uploading and distribution of this book via the Internet or via any other means without the permission of the author is illegal and punishable by law. Thank you for purchasing only authorized electronic editions, and for withdrawing your consent or approval for electronic piracy of copyrightable materials. Your support of the author's rights, as well as your own integrity is appreciated.

Pierucci Publishing books may be purchased in bulk at special discounts for sales promotion, corporate gifts, fund-raising, or educational purposes. Special editions can be created to specifications. For details, contact the Special Sales Department, Pierucci Publishing, PO Box 2074, Carbondale, CO 81623 or Publishing@PierucciPublishing.com or toll-free telephone at 1-855-720-1111.

"Climate change is a tool that is being used against the West, our freedom, and our future. The climate change narrative is built on misinformation, disinformation, and manipulated data. It is fostered by deliberate propaganda, indoctrination, and censorship to drive horrendous energy policy and the eventual loss of freedom. Western Civilization is at risk from climate policies, not climate change!"
-Frank Lasèe

TABLE OF CONTENTS

Introduction: The World Is Not Ending
   The Earth Is Warming
   CO_2 Is Not the Bad Guy
   We've Been Lied to About Renewable Energy
   You Can Be a Climate "Denier" and Still Save the Earth
   Central Planning Is the Path to Hell on Earth

PART ONE: THE CULT OF CLIMATE CHANGE

Chapter One: The Climate Change Narrative is a Lie
   People and the Planet Thrive on Free-Markets

Chapter Two: Literal and Figurative Energy Crises

Chapter Three: Climate Change is A Religion
   The Doctrine of the Cult of Climate Change
   Beliefs & Commandments of the Cult of Climate Change

Chapter Four: Becoming A Heretic

Chapter Five: Unveiling Lies in The Media
   The "97% of Scientists Agree" Lie
   Censorship
   Galileo Galilei
   Trofim Lysenko
   António Egas Moniz
   Cocaine, The Wonder Drug?

PART TWO: CO_2 LIES

Chapter Six: The CO_2 Lie
   What is "the Greenhouse Effect?"
   Greenhouse Gases Matter

Chapter Seven: CO_2: The Miracle Molecule
   CO2: The Catalyst for Crop Growth

Due to More CO_2 Food Crop Yields Are on the Rise

PART THREE: CLIMATE LIES

Chapter Eight: The Carbon Cycle Unveiled
Understanding the IPCC and Climate Science
Testimonies from the 2007 U.S. Senate Environment and Public Works Committee Minority Staff Report

Chapter Nine: Climate Models Aren't Ready for Prime Time
Global Temperatures

Chapter Ten: Climate Fake News
Climate Change is Local, Not Planet-Wide...
Temperature Isn't as Accurate as They Are Telling You
The Urban Heat Island Effect

Chapter Eleven: Altered Data & Misinformation
Why Do NOAA and NASA Lie?
Blatant Data Tampering
Covering Up Cooling Trends
ClimateGate
Raw Data Adjustments
NOAA Falsified Congo Data in Map of The World

Chapter Twelve: It's Not as Hot as It Used to Be
Record Heat in 1936
The Polar Vortex

PART FOUR: DEBUNKING OCEAN LIES

Chapter Thirteen: Changing Sea Levels
Where Is the Water?

Chapter Fourteen: Coral Reefs and Island Expansion
Coral Reefs Are Doing Just Fine
The Truth About Ocean Islands
Evidence of Island Expansion
The Truth About Ocean pH and Acidification

Chapter Fifteen: Glaciers & Natural Cycles

Chapter Sixteen: Melting Glacier Lies
Polar Bears Are Doing Just Fine – Better than Fine

Chapter Seventeen: Antarctica Has Had No Melting in 70 Years or More

Chapter Eighteen: How Earth's Weather is Made
　Weather & The CO_2 Lie
　Hurricanes
　Normalized U.S. hurricane damage from 1900–2022
　Tornadoes
　Droughts
　Floods
　Fires, Fire Propaganda, and Statistics

PART FIVE: THE WORLD IS SAFER THAN EVER

Chapter Nineteen: Organisms Including Humans Are Thriving
　Less Disease with Better Living Conditions
　Most Animals and Insects Are Doing Fine There Is No Mass Extinction
　There's Less Poverty in the World Now than Ever Before
　All Is Well with the Climate - Climate Policy is the Real Danger

PART SIX: ENERGY

Chapter Twenty: It Used to be Dark – It Still Is in Many Places
　Abundant Affordable Reliable Energy Matters
　China
　India
　Coal
　Oil
　Products Made from Oil, Coal, and Natural Gas
　Natural Gas (Also Known as Methane)
　Electric Grid Fundamentals
　Adding Wind and Solar Makes Your Electric Bill Go UP, UP, UP!
　Green Hydrogen: Another Boondoggle of Wishful Thinking
　Electric Car Batteries
　ESG
　Conclusion

INTRODUCTION:
THE WORLD IS NOT ENDING

> "There is no observational evidence that the addition of anthropogenic (caused by humans) greenhouse gas emissions have caused any temperature perturbations in the atmosphere."
> Dr. George T. Wolff [1,2,3]
> Award-Winning Atmospheric Scientist, and Former Member of the EPA's Science Advisory Board
> From *The New York Times* Book Review, July 18, 1976

Climate change is a threat to our economy, our families, and our very human existence... but in not the way you might think.

Having served as a Senator for the great State of Wisconsin, I have written this book from the perspective of someone who is interested in separating the truth from the lies about climate change, and the energy policies it drives. Also, as someone who understands why we are being fed those lies, I'm uniquely qualified to serve as your guide to understanding this complex hot topic, the heat from which has nothing to do with rising temperatures.

I've intentionally written this book in such a way that your average high school freshman will be able to follow along, and by reading this book, have more common-sense wisdom and factual evidence than most, if not all of their teachers and future professors.

---

[1] *Renowned for his expertise, Dr. George T. Wolff served on a committee of the National Oceanic and Atmospheric Administration (NOAA) and has authored over 100 peer-reviewed studies. His international recognition stems from his profound knowledge of air quality and the intricate relationships between meteorology and air quality.*

[2] https://www.epw.senate.gov/public/index.cfm/press-releases-all?ID=10fe77b0-802a-23ad-4df1-fc38ed4f85e3

[3] http://www.airimprovement.com/personnel/dr_george_t_wolff_principal.html

Climate and Energy Lies will equip readers with a blueprint for affecting change in their communities by becoming beacons of truth in an environment plagued not by rising temperatures, but by rising propaganda around climate change; which has been deliberately designed to impoverish, starve, and force us to submit to globalist and government control.

Globalists have cleverly manipulated three aspects of a complex system – climate, weather, and temperatures – into simplified propagandist messages in order to lead the world down a path to poverty, food shortages, and ultimate control over the masses.

Our understanding of the Earth's climate evolved over the centuries of recorded weather patterns, temperatures, and other somewhat complex factors into the political concept of climate change. Sadly, however, with ample evidence that climate change is not an urgent issue or concern, it has nevertheless been wielded as a weapon in today's media, economic, social, and especially political landscapes.

In the 1970s, the prevailing concern was of the coming Ice Age. In fact, an infamous TV special hosted by Leonard Nimoy of *Star Trek* fame was dedicated to this very concern; it was called "The Coming Ice Age.[4]" The mainstream media's supposed scientific experts, high-profile influencers such as Nimoy, and published studies from that era warned about this looming "Ice Age" due to a noticeable drop in temperatures that was recorded between 1940

> The New York Times Book Review / July 18, 1976
>
> **The Cooling**
>
> So writes Stephen Schneider, a young climatologist at the National Center for Atmospheric Research in Boulder, Colo., reflecting the consensus of the climatological community in his new book, "The Genesis Strategy." His warning, that present world food reserves are an insufficient hedge against future famines, has been heard among the scientific community for years—for example, it was a conclusion of a 1975 National Academy of Sciences report. But Schneider has decided to explain the entire problem, as responsibly and accurately as he can, to the general public, and thus has put together a useful and important book.
>
> Schneider quotes University of Wisconsin climatologist Reid Bryson as saying that 1930-1960 "was the most abnormal period in a thousand years—abnormally mild." In fact, conditions of steady, warm weather in the northern hemisphere during that time favored bumper harvests in the United States, the Soviet Union, and the wheat belt of northern India and Pakistan. In 1974 Schneider and Bryson tried to explain to a White House policy-making group why conditions are likely to worsen. One of the most depressing anecdotes in the book is Schneider's description of the deaf ear their warnings received.

---

4   https://www.imdb.com/title/tt0894213/

to 1979. Unfortunately, most climate change proponents and personalities today rarely refer to this recent historical event. In fact, like so much other climate "history," they pretend as if it was only a tiny minority of people and beliefs of the late 1970s, or that it wasn't true at all.

Not only do we have societal amnesia regarding the mixed messages around the Earth's ever-changing temperatures, they also have cooked the books. You see, the National Oceanic and Atmospheric Administration (NOAA) and National Aeronautics and Space Administration (NASA) have mysteriously altered temperature records, even having largely *erased* the cooling trend from 1940 to 1979, a time during which carbon dioxide levels were rising. These adjustments have been made to edit history to support today's "global warming narrative;" now more commonly called the "climate change narrative." I'll provide evidence in this book of other instances of nefarious tampering of information, in an effort to provide you with the actual facts.

It is my opinion, as well as that of many experts with far more experience and depth of knowledge than Greta Thunberg or anybody in Congress, that the Earth's continually changing temperature, climate, and weather is a natural, ongoing process over which we have no control. These things have always changed...and always will.

This article from *Newsweek* magazine on April 28, 1975, clearly shows the cooling trend from 1940 to 1975. Later, I will show you the NOAA records where this and many other records have been altered because they don't fit the narrative.

SCIENCE

The Cooling World

There are ominous signs that the earth's weather patterns have begun to change dramatically and that these changes may portend a drastic decline in food production—with serious political implications for just about every nation on earth. The drop in food output could begin quite soon, perhaps only ten years from now. The regions destined to feel its impact are the great wheat-producing lands of Canada and the U.S.S.R. in the north, along with a number of marginally self-sufficient tropical areas—parts of India, Pakistan, Bangladesh, Indochina and Indonesia—where the growing season is dependent upon the rains brought by the monsoon.

The evidence in support of these predictions has now begun to accumulate so massively that meteorologists are hard-pressed to keep up with it. In England, farmers have seen their growing season decline by about two weeks since 1950, with a resultant over-all loss in grain production estimated at up to 100,000 tons annually. During the same time, the average temperature around the equator has risen by a fraction of a degree—a fraction that in some areas can mean drought and desolation. Last April, in the most devastating outbreak of tornadoes ever recorded, 148 twisters killed more than 300 people and caused half a billion dollars' worth of damage in thirteen U.S. states.

Trend: To scientists, these seemingly disparate incidents represent the advance signs of fundamental changes in the world's weather. The central fact is that after three quarters of a century of extraordinarily mild conditions, the earth's climate seems to be cooling down. Meteorologists disagree about the cause and extent of the cooling trend, as well as over its specific impact on local weather conditions. But they are almost unanimous in the view that the trend will reduce agricultural productivity for the rest of the century. If the climatic change is as profound as some of the pessimists fear, the resulting famines could be catastrophic. "A major climatic change would force economic and social adjustments on a worldwide scale," warns a recent report by the National Academy of Sciences, "because the global patterns of food production and population that have evolved are implicitly dependent on the climate of the present century."

A survey completed last year by Dr. Murray Mitchell of the National Oceanic and Atmospheric Administration reveals a drop of half a degree in average ground temperatures in the Northern Hemisphere between 1945 and 1968. According to George Kukla of Columbia University, satellite photos indicated a sudden, large increase in Northern Hemisphere snow cover in the winter of 1971-72. And a study released last month by two NOAA scientists notes that the amount of sunshine reaching the ground in the continental U.S. diminished by 1.3 per cent between 1964 and 1972.

To the layman, the relatively small changes in temperature and sunshine can be highly misleading. Reid Bryson of the University of Wisconsin points out that the earth's average temperature during the great Ice Ages was only about 7 degrees lower than during its warmest eras—and that the present decline has taken the planet about a sixth of the way toward the Ice Age average. Others regard the cooling as a reversion to the "little ice age" conditions that brought bitter winters to much of Europe and northern America between 1600 and 1900—years when the Thames used to freeze so solidly that Londoners roasted oxen on the ice and when iceboats sailed the Hudson River almost as far south as New York City.

Just what causes the onset of major and minor ice ages remains a mystery. "Our knowledge of the mechanisms of climatic change is at least as fragmentary as our data," concedes the National Academy of Sciences report. "Not only are the basic scientific questions largely unanswered, but in many cases we do not yet know enough to pose the key questions."

Extremes: Meteorologists think that they can forecast the short-term results of the return to the norm of the last century. They begin by noting the slight drop in over-all temperature that produces large numbers of pressure centers in the upper atmosphere. These break up the smooth flow of westerly winds over temperate areas. The stagnant air produced in this way causes an increase in extremes of local weather such as droughts, floods, extended dry spells, long freezes, delayed monsoons and even local temperature increases—all of which have a direct impact on food supplies.

"The world's food-producing system," warns Dr. James D. McQuigg of NOAA's Center for Climatic and Environmental Assessment, "is much more sensitive to the weather variable than it was even five years ago." Furthermore, the growth of world population and creation of new national boundaries make it impossible for starving peoples to migrate from their devastated fields, as they did during past famines.

Climatologists are pessimistic that political leaders will take any positive action to compensate for the climatic change, or even to allay its effects. They concede that some of the more spectacular solutions proposed, such as melting the arctic ice cap by covering it with black soot or diverting arctic rivers, might create problems far greater than those they solve. But the scientists see few signs that government leaders anywhere are even prepared to take the simple measures of stockpiling food or of introducing the variables of climatic uncertainty into economic projections of future food supplies. The longer the planners delay, the more difficult will they find it to cope with climatic change once the results become grim reality.

—PETER GWYNNE with bureau reports

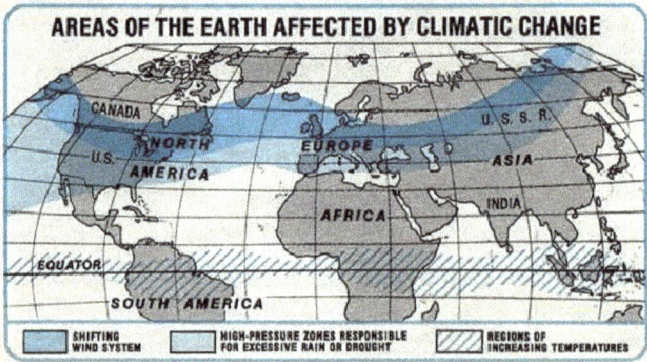

AREAS OF THE EARTH AFFECTED BY CLIMATIC CHANGE

SHIFTING WIND SYSTEM — HIGH-PRESSURE ZONES RESPONSIBLE FOR EXCESSIVE RAIN OR DROUGHT — REGIONS OF INCREASING TEMPERATURES

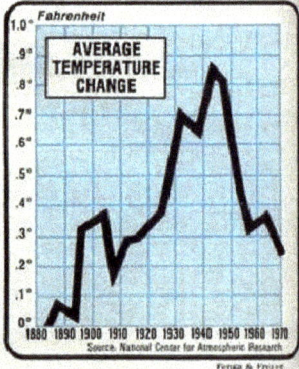

AVERAGE TEMPERATURE CHANGE

Source: National Center for Atmospheric Research

Newsweek, 28 April 1975

In this book, I will challenge nearly all of the deeply held beliefs of the climate change narrative. To some, this investigation is considered heresy. Nevertheless, this book will call into question the honesty, integrity, and accuracy of records that are being used to promote the climate change narrative today; supported with ample evidence, of course. In its original state, this manuscript was called, *Confessions of A Climate Change Heretic* because challenging climate change in today's volatile environment is, frankly, a little like challenging a religion. Do you think I'm being melodramatic? You be the judge...

Want to fight climate change? Have fewer children

"*The Guardian*" 12 July 2017

CLIMATE NOTE · Jul 25, 2023

The prevalence of Climate Change Psychological Distress among American adults

"*Yale.edu*" 25 July 2023

SCIENCE

Your Crushing Anxiety About the Climate Crisis Is Normal

"*Smithsonianmag.com*" 18 May 2022

Climate depression is real. And it is spreading fast among our youth

The Guardian" 4 November 2021

Study: Climate change deniers changing tactics to spread misinformation
"CNN" 17 Jan 2024

According to these headlines, we're either on one end of the spectrum, having fewer children, experiencing more anxiety and depression, and frightening our loved ones about climate change. Or, if you're living in peace about the Earth's continually changing temperatures, climate, and weather patterns, well you're a dangerous misinformation spreader and "climate denier." See the discrepancy?

It was reported by Pew Research in 2022 that among nineteen countries, climate change supersedes worry about cyber-attacks, the global economy, and even infectious disease[5]. Among those who are psychologically distraught over climate change, it is my suspicion that most of them have never actually delved into the scientific details about the Earth's climate, temperatures, or weather patterns.

In a page out of some Orwellian literature, slogans such as "it's the consensus" or "97% of scientists agree" have become the rallying cry of "climate changers." Dissenting voices, which I referred to above as "climate deniers," are considered heretics.

However, what's heretical is the fact that climate change is being weaponized against the people. It extends far beyond environmental concerns; their climate policies hurt our energy supply, increase energy costs, and diminish electricity reliability and availability. In several countries, including the U.S., they have the manipulation of the agricultural industry that grows our food clearly in their sights. They are working on limiting food supply through regulations, and changing what you eat "for the climate." The intricate web of motivations (and lies) behind climate change activism and the secret behind-the-scenes agenda of the rich and powerful will be revealed throughout this book – and it will shock you.

It boils down to control, greed, and power aimed at reshaping the Western World and ending capitalism as we know it. The Great Reset, as they have been openly calling it.

---

[5] https://www.pewresearch.org/global/2022/08/31/climate-change-remains-top-global-threat-across-19-country-survey/

Having spent nearly twenty years in politics as an elected State Representative and Senator in Wisconsin, I know that "following the money" is just one of the avenues to discern what the powerful are really after. Most of the climate change evildoers already have plenty of money; it's power over the masses that they are seeking now. In short, they are looking to grow their power and influence at the expense of yours and mine.

Climate change is being sold as a man-made catastrophe—it is not, which I'll show you clearly in the following pages. We are being propagandized with a purpose; it is a large-scale global gaslighting operation that blames innocent people for a problem which does not truly exist.

Among many messages utilized in the climate change propaganda and global gaslighting is that the Earth is overpopulated, which justifies tertiary agendas such as rampant funding and celebration of abortion. Falling birth rates will potentially cause greater destruction for humanity than so-called climate change.

As a result of overpopulation lies, more people are better fed than ever before; politicians will argue that Westerners, who live too well and use too many resources, should be on a more level playing field with the developing world. These politicians believe Westerners should have, do, and consume less, arguing that the only way to reduce CO_2 emissions is by doing less, consuming less, traveling less, and eating less. They continue to tell us lies about energy and hide the human suffering it will cause.

What I see isn't that the West will be on a more level playing field, but that we will simply be leveled. We are being deceived into surrendering our prosperity and freedoms. The globalists are working to get us in a place where we are powerless to fight back. Their agenda isn't to raise our standard of living even more, nor raise the living standards of those who don't have our Western standards. They are working to lower ours while keeping Africans from rapidly increasing theirs.

This ideology has given rise to a multi-trillion-dollar Climate Change Industrial Complex wherein taxpayers are required to subsidize an unnecessary "energy transition," and endure endless restrictions and regulations for a supposed man-made climate catastrophe. Among these subsidies are zero-carbon initiatives and even sustainable developments; the likes of which are planned for, among many other places, the west side of

Maui where rampant fires conveniently leveled Lahaina in 2023. The State of Hawaii has indicated "climate change" as part of their reason for proposing Senate Bill 3381, which declares eminent domain over nearly the entire west side of the island of Maui. The State wants to take total control of this area.

After years of thorough study, immersion into "real" scientific research, listening to experts, and exploring both sides of the climate change debate, my beliefs have crystallized into the various arguments we will go over in this book. We'll review these in a manner that anyone can understand without having a background in physics, meteorology, or even a high school diploma.

THE EARTH IS WARMING

Since about 1850, the world has been naturally warming, following the "Little Ice Age" from about 1350–1850, also known as "preindustrial times." Preindustrial times make the sale of climate change easier. A little warmer than the Little Ice Age doesn't sound nearly as scary or desirable as the more romantic preindustrial times.

Future warming has many positive aspects, including longer growing seasons. A gradually warming world has no inherent downsides. Yet, in the upside-down world of climate propaganda, we are told the opposite on a daily basis. It is even worked into TV shows and sports articles, somehow.

CO_2 IS NOT THE BAD GUY

CO_2 is not the sole control knob for the climate or Earth's temperature. In fact, much of the reported warming is the result of NOAA and NASA adjusting data, and making up data from 70, 100, and 140 years ago. They have made it up because it simply did not exist. They then pretend this made-up data is highly accurate; we'll detail several examples of this made-up information in the pages that follow.

CO_2 is a substantial benefit to the world, promoting plant and phytoplankton growth, aiding in forest regeneration, nourishing the soil, and sustaining life.

Climate zealots are focused on net-zero, which means reducing CO_2 emissions, to none, zero, zip, nada. The climate purists or zealots want no CO_2 emissions at all. This will allow for initiatives to dominate which have never worked at scale, such as Carbon Capture and Storage (CCS). If CCS does work, which has not been shown, it will make our energy far more expensive, which will make *everything* far more expensive.

Instead, shouldn't we direct our efforts toward improving the world and preparing for potential changes due to warming? Adaptation is far better than wasting our money unnecessarily on mitigation, or trying to stop releasing CO_2 at all.

WE'VE BEEN LIED TO ABOUT RENEWABLE ENERGY

Fossil fuels, coal, oil, and natural gas provide four fifths or 80 percent of the world's energy. Totalitarian communist China–using climate policy to its advantage–poses a substantial threat to global freedom and security. China uses more than half of all the coal in the world, which provides more than half of all their energy, not just electricity. China is the world's biggest emitter of CO_2, more than all industrial nations combined.

Wind and solar energy are not the ultimate solution, as they take up ghastly amounts of land. Wind towers and solar panels also require huge amounts of fossil fuels to manufacture and transport! They have a negative effect on the environment and they're plainly unreliable.

If we want the luxury of a full-time electric grid, something we often take for granted, we cannot rely on wind towers and solar panels, nor electric cars. The products made by and with energy from fossil fuels don't just improve our lives, they also save lives.

YOU CAN BE A CLIMATE "DENIER" AND STILL SAVE THE EARTH

The preservation of the environment, wildlife, and valuable croplands should take precedence over the installation of wind towers and solar panels. I will show you how the expansion of the use of fossil fuels has benefited the environment, as well as the rest of us.

Later, I will show you this in the case of the Dominican Republic and Haiti, each occupying one half of the Caribbean Island of Hispaniola. Fossil fuels have provided measurable economic, environmental, and political benefits. As of this writing, Haiti has descended into chaos, which I believe has a lot to do with their lack of energy and all of the stability it brings.

Using more fossil fuels, not less, is good for the environment, and I will show you why.

CENTRAL PLANNING IS THE PATH TO HELL ON EARTH

Central planning often leads to shortages and disastrous outcomes. History is filled with the failures of central planning, and our elected leaders often do not know best. They are subject to many pressures and often graft. They often pick wrong when betting on the next great technological breakthrough. They are spending other people's money, so they are not held accountable for their mistakes.

Market discipline holds people accountable if consumers don't want it or it doesn't have enough perceived value for the price. Whatever product or service just will not be produced because there aren't enough buyers. In the government world of subsidies, this doesn't hold true because the government picks up a large part of the tab. Wind, solar, and electric vehicles (EV) are hugely subsidized with taxpayer money that is borrowed and added to our exploding $34 trillion national debt.

The climate discourse has been manipulated for financial gain, power, and control. The climate change minions are being used by globalist individuals, corporations, and politicians to increase their power, income, and control. The United Nations (UN) and the World Economic Forum (WEF) are the most visible and assertive villains interested in weaponizing climate change to gain total domination over businesses, economies, real estate, the food supply, and your own personal freedom.

Currently, the people in Europe are doing a much better job than the United States at pushing back against the climate agenda; this is mainly led by farmers. Granted, Europe is further down the path of climate policy self-destruction. In the West, most politicians, even some so-called conservatives, adhere to the climate change narrative and the policies being weaponized to ensure your compliance. These policies lead to high energy and food prices.

Watch *No Farmers, No Food*, a documentary that interviews European farmers who face the threat of being put out of business. It reveals the agenda to make bugs our primary protein source, which is being pushed by the WEF and UN, as well as other climate propagandists who also push vegan agendas.

This book is not just a call to action, it is a vital warning. I urge readers to critically examine the motives behind climate change activism from the obvious money and land grabs to the ultimate question of power: who wants more of it and how are they going to get it?

As history has taught us, when propaganda is used to hide scientific inquiry and condemn critical thinkers, this divides society, and usually it's for the purpose of controlling the population for nefarious purposes. It is for this reason that I feel it is my duty to warn the public and expose the truth. We cannot continue to shrug off the climate change debate and allow critical thinkers to be labeled "climate deniers." History shows us that complying with this tactic of bullying, censorship, and name-calling will lead us down a very scary road—one that dismantles our families, our finances, and our freedoms.

Part One:
THE CULT OF CLIMATE CHANGE

CHAPTER ONE:
THE CLIMATE CHANGE NARRATIVE IS A LIE

"Half of humans around the world think that climate change could make humans go extinct. There's zero science to support that. There's not even very much science fiction to support that."
Michael Shellenberger

My unique perspective and sense of authority on this topic is derived from my time as State Senator in Wisconsin and President of the Truth in Energy and Climate Organization (www.truthinenergyandclimate.com). Right now, critical thinking and common sense are lost in society, and particularly in politics.

In the administration of Former Governor Scott Walker, I was responsible for Workers Compensation (Wisconsin has one of the best and most fair systems in the nation), which gave me a special heart for the needs of Americans who are injured at work. Government service compelled me to look deeper into how governments work to either help and protect, or harm and subjugate people. What I learned is that protecting people sometimes feels like sledding uphill.

Witnessing the constant bureaucracy and competing special interests that hinder or drive political action gave me a clear perspective on how governments and corporations often work in tandem to manipulate people for their own benefit. This has metastasized into a deadly societal cancer permeating climate politics, propaganda, and policy.

The Wisconsin district I represented included power facilities for electricity, that included coal, bio-gas, biodigesters, natural gas, two nuclear power plants, wind towers, and now, a pretty large taxpayer-subsidized solar farm.

The reason climate change has focused on control of food and energy is that those who seek control over us understand that controlling energy and food is the best method to control everyone and everything. Without energy and food, we die. With limited food and energy, we lead limited lives. And we will spend the majority of our time focused on limited, expensive energy and food supplies.

To control their production to limit CO_2, methane, and nitrous oxide means they will micromanage every aspect of our lives. It will become a requirement, because life as we know it requires energy and food. And energy and food release CO_2. Even the so-called "green" sources release CO_2. This control is what the global elite and corrupt politicians are working to get. This is wrong, wrong, and wrong.

My wife Amy Joy and I have a blended family with seven children (only the youngest being a boy) and two grandchildren. When I think about those elites getting control over us, our children, and our grandchildren's future, I am motivated to fight for the prosperity of my family and yours. We are in this together. Governments are *supposed* to help people and enhance our lives, not seek to control every aspect of them. Not even if it's "for our own good, to save the climate" (that doesn't even need saving).

Our government should have limits and help, rather than harm, us. It should serve us, not boss us around and rule over us. It should enhance freedom and protect us from each other, not seek to micromanage our lives.

It shouldn't force EVs on us against our will, subsidize them, and take away our choice of owning gas cars, if we want them and can afford them. Nor should it add to our $34 trillion dollar debt subsidizing wind and solar, mainly purchased from totalitarian China. This makes no sense! More on this later in the book.

PEOPLE AND THE PLANET THRIVE ON FREE-MARKETS

I'm an advocate for free-market capitalism, which means that free people should willingly engage in free markets with environmental protection laws to reasonably protect our land, air, and water. And we should have laws to protect the weak from the strong. The rule of law, enforcement of private contracts, and private property rights are all things we must preserve and enhance to continue to be free and prosperous people. And of course, we must maintain our freedom of speech and thought.

When our government engages in central planning, it is interrupting free markets. It is telling us, through subsidies and regulations, what the market and we, the people, should be doing. As history shows us, a lot can go wrong with this model. These government central planning measures tend to fail because they often base their decisions on who they know, and who will benefit. When wrong bets are made, they must assert increased control over people which creates an inefficient stagnancy to our economies. Free markets respond flexibly, while controlling governments double down on the failure.

Competition means that businesses or providers work harder to provide a better product at a cheaper price; that's one way capitalism creates a better society for us all. Privatized businesses–not governments–are competing to truly satisfy their customers so they can grow their market share and prosper. If the government doesn't perform well or provide an excellent product, there's little to no accountability; because you don't have a choice, there is no competition. They still get to tax you for "their product" and if you don't participate, you can be put in prison.

The United States hasn't been a great ambassador for free-market capitalism. Our leaders focused on democracy instead. But because they had the wrong idea that every nation out there was ready for "democracy," many aren't.

Many people in the world have no idea how an economy works or what to do with this newfound power to the people, that we call "democracy." It's especially confounding in countries where people don't have a history of capitalism, individual freedom, private property, and the rule of law; and where the vast majority have little or no education or experience running large organizations…let alone a country.

The United States is actually a constitutional republic, not a pure democracy. Our focus should be on protecting individual rights and limiting government power; holding the government accountable to serving citizens, not controlling them. More for the regular guy and far less for the benefit of government and private sector elites. It's been virtually inverted from this original premise.

There's a common misconception that Scandinavian countries are socialist due to their well-developed welfare systems and high taxes. However, the Scandinavian countries actually rate higher on the Heritage Foundation's Economic Freedom Index than the United States. They embrace a high level of government benefits, but they aren't crippled with as much centralized planning or bureaucracies as in the U.S. These important concepts are introduced here, because the climate change narrative and the policies they drive are an attack on free market capitalism. These climate policies are the road to serfdom.

Energy plays a crucial role in fostering environmental care and improving all of our lives. Globally, there are billions of people who lack access to sufficient energy. As energy access improves, so do lives. As lives improve, people change their focus from hand-to-mouth subsistence to prosperity and, in turn, environmental stewardship.

Despite the unfortunate turn in the U.S. and globally towards central planning and government overreach, I'm optimistic about the future. The governmental push for net-zero decarbonization (deCO$_2$ing everything) is a threat to freedom, as is pressure on our food supply through the regulation of the tiny trace gasses, like methane, and nitrous oxide. Atmospheric Physicist Dr. Richard Lindzen, at Massachusetts Institute of Technology (MIT), tells us that "controlling carbon (CO$_2$) is a bureaucrat's dream. If you control carbon, you control life." The climate cult, not satisfied with CO$_2$, added methane and nitrous oxide to their ever-growing list of things to control.

We may be able to avoid the coming climate change driven economy of despair and return to a more free-market energy economy that evolves more dynamically than the pace of law and promotes prosperity that begets environmental stewardship. The choice of who we elect to make our laws is more important than ever. Those who seek total control have found the tool of manmade climate change; and they have it in a vice grip.

CHAPTER TWO:
LITERAL AND FIGURATIVE ENERGY CRISES

"I don't believe there is a climate crisis… The world we live in today is filled with misinformation. It is up to each of you to serve as judges, distinguishing truth from falsehood based on accurate observations of phenomena."
John Francis Clauser[6]

In a later chapter, I'll describe how climate change has become like a religion. A no-win religion. In the cult of climate change all non-believers who speak up are labeled "deniers." And similar to witches during the Salem Witch Trials, they're going to be figuratively tied up and thrown in the river. In the cult of climate change, the narrative relies on the presupposition that our weather and climate are getting worse due to CO_2 emissions from fossil fuels such as oil, coal, and natural gas. The religion of climate change preaches fear: the fear of climate Hell. Hell equates to the imminent collapse of human civilization and the death of not millions, but billions of people.

Do you want salvation? Well then, you're going to have to subscribe to our agenda: pay your tithe and engage in some good old-fashioned flogging and repentance. If you don't want to fight to end fossil fuels and replace them with wind, solar, and batteries, you're a sinner and you're leading others to hell alongside you.

---

[6] **John Francis Clauser**, an esteemed American theoretical and experimental physicist, was awarded the 2022 Nobel Prize in Physics jointly with Alain Aspect and Anton Zeilinger for experiments with entangled photons, establishing the violation of Bell inequalities, and pioneering quantum information science. Clauser is renowned for his contributions to the foundations of quantum mechanics, particularly the Clauser–Horne–Shimony–Holt inequality.

In truth, CO_2-emitting energy sources, which provide 83% of the world's energy, haven't changed much in the past fifty years. Despite all the clamoring about overpopulation, wasteful energy usage, and, of course, spending trillions of dollars on wind turbines and solar panels, we have little to show for it.

Without coal, natural gas, and oil, along with the fertilizers they provide, billions of people will die from starvation. Hindering production of coal, natural gas, and oil drives up their costs and holds back our ability to make critical advancements to a world that needs to continue to improve.

In the coming decades, government interference with fossil fuel mining poses a threat to our entire economy. Subsidies for wind, solar, batteries, rare metal mining, and hydrogen designed to reduce CO_2 are increasing the U.S. debt, which is already at alarming levels. Threatening a total economic collapse in the coming years...

You'll learn through this book that we have nothing to fear by fluctuating temperatures, weather patterns, and climates. We do have the very real threat of poverty, rising energy costs, and food supply shortages due to climate change *policies* fostered by propaganda and "carbon-neutral" mandates, subsidies, and alternative energy initiatives.

Energy is everything, and that isn't merely a spiritual mantra you heard in yoga class or in some self-help seminar. Everything around us has to do with energy, literally as well as figuratively. When you walk, you are required to use energy that you've ingested through calories, which you receive through eating, which is at the center of most human social interaction. When you walk, you create CO_2 with your breath, and even when you sleep you need energy, and you exhale about two pounds of CO_2 a day.

An ever-growing energy supply is also required to power the internet, AI, cryptocurrencies, more EVs, and the various luxuries you enjoy in your home, from bathing to heating the house in the winter. Fossil fuels are critical in development, and powering life-saving medical equipment. All cars also require fossil fuels to manufacture; even electric ones.

Plainly, the idea that we must eliminate fossil fuels immediately is naive. At best, it only leads to economic harm. At worst, it leads to the death of the majority of humans on Earth. Without fossil fuels, we all return to subsistence farming. Except for maybe the elites, who surely have plans to exempt themselves from this evil fate of hardship and death.

If I believed in the climate change fear campaigns and the idea that we are destroying Earth with fossil fuels, eating meat, and releasing CO_2, I'd be a devotee of the climate change cult too. This book will give you the facts to see it is a false religion.

I understand that many of my readers are truly concerned about the Earth and our use of its resources. You might be passionate about nature preservation. Well, I'm with you.

In today's world, the battle is not between left and right or red and blue. Today's battle is between totalitarian control and individual freedom. Instead of fighting among ourselves, we need to aim at the nefarious globalists and their often-misinformed minions who want to separate us from freedom and prosperity. By hiding their own environmentally-damaging corporate practices while gaslighting society.

"You know," a lifelong Hawaiian and political leader said staring my friend dead in the eyes, tears streaming down his cheeks, "we caused the Lahaina fires. We caused them with our plastic straws and our wasteful energy usage."

Don't be like this man who has been propagandized into explaining away the trauma of one of the largest disasters in U.S. history with climate change hogwash. It is easy to blame climate change, rather than taking responsibility for not managing the brush so it doesn't burn or not delivering water to put out the fire in time.

Richard Lindzen[7,8] states,

Future generations will wonder in bemused amazement that the early 21st century's developed world went into hysterical panic over a globally averaged temperature increase of a few tenths of a degree, and, on the basis of gross

---

[7] **Richard Lindzen**, an esteemed American atmospheric physicist with a Harvard education and a prolific publication record, boasting over 200 scientific papers and books. Lindzen held the prestigious position of Alfred P. Sloan Professor of Meteorology at the Massachusetts Institute of Technology (MIT). His contributions to the field of atmospheric physics have left an indelible mark, establishing him as a respected figure in climate science. Lindzen's expertise and academic achievements reflect a dedication to advancing our understanding of meteorology and its implications.

[8] https://www.aei.org/carpe-diem/quotations-of-the-day-on-climate-alarmism/

> *exaggerations of highly uncertain computer projections combined into implausible chains of inference, proceeded to contemplate a roll-back of the industrial age.*

On another level, climate lies threaten the modern age where we, the common man, live longer and better, enjoy less disease, fewer natural disasters, and more prosperity than all of recorded history. Climate lies will inevitably lead to slavery and a complete reversal of the progress we have made.

For example, the global elite are seeking to eliminate or greatly limit most people from eating meat. Most of us don't want to do what they are advocating, which is to eat bugs instead of meat. Yes, the UN and WEF are pushing bug eating! Tyson Foods is building a major insect-for-food processing facility in the U.S. Insects are approved to be added to our food supply in the U.S. and Europe. People were not meant to eat insects as a regular part of our diet on a continual basis.

This book is my rescue operation for a society that's unknowingly slipping into slavery through propaganda and climate mind control. Before we upend our economy and surrender our God-given rights and freedoms, I pray this book will help break the spell of the climate cult that's being spread by propaganda and coercion.

CHAPTER THREE:
CLIMATE CHANGE IS A RELIGION

> *"I am a skeptic...Global warming has become a new religion."*
> Dr. Ivar Giaever

Dr. Ivar Giaever is a Norwegian-American engineer and physicist who shared a Nobel Prize in Physics in 1973 with Leo Esaki and Brian Josephson for their discoveries regarding tunneling phenomena in superconductors.

In 2011, Dr. Giaever resigned from his position with the American Physical Society (APS) over its official position on climate change and global warming fears. In a resignation letter to APS Executive Director Kate Kirby, Dr. Giaever tells them he "could not live with" the APS statement on global warming, which reads:

> *The evidence is incontrovertible: Global warming is occurring. If no mitigating actions are taken, significant disruptions in the Earth's physical and ecological systems, social systems, security, and human health are likely to occur. We must reduce emissions of greenhouse gasses beginning now.*

Dr. Giaever elaborates in the email to Kirby:

> *In the APS it is ok to discuss whether the mass of the proton changes over time and how a multi-universe behaves, but the evidence of global warming is incontrovertible? The claim (how can you measure the average temperature of the whole Earth for a whole year?) is that the temperature has changed from ~288.0 to ~288.8-degree Kelvin in about 150 years, which (if true) means to*

> *me is that the temperature has been amazingly stable, and both human health and happiness have definitely improved in this 'warming' period.[9]"*

From: Ivar Giaever [mailto:giaever@XXXX.com]
Sent: Tuesday, September 13, 2011 3:42 PM
To: kirby@aps.org
Cc: Robert H. Austin; 'William Happer'; 'Larry Gould'; 'S. Fred Singer'; Roger Cohen
Subject: I resign from APS

Dear Ms. Kirby

Thank you for your letter inquiring about my membership. I did not renew it because I can not live with the statement below:

Emissions of greenhouse gases from human activities are changing the atmosphere in ways that affect the Earth's climate. Greenhouse gases include carbon dioxide as well as methane, nitrous oxide and other gases. They are emitted from fossil fuel combustion and a range of industrial and agricultural processes.
The evidence is incontrovertible: Global warming is occurring.
If no mitigating actions are taken, significant disruptions in the Earth's physical and ecological systems, social systems, security and human health are likely to occur. We must reduce emissions of greenhouse gases beginning now.

In the APS it is ok to discuss whether the mass of the proton changes over time and how a multi-universe behaves, but the evidence of global warming is incontrovertible? The claim (how can you measure the average temperature of the whole earth for a whole year?) is that the temperature has changed from ~288.0 to ~288.8 degree Kelvin in about 150 years, which (if true) means to me is that the temperature has been amazingly stable, and both human health and happiness have definitely improved in this 'warming' period.

Best regards,

Ivar Giaever

Nobel Laureate 1973

September 13, 2011 from Nobel Prize Winner and Physicist Dr. Ivar Giaever

One reason I often refer to climate change as a religion is because the vast majority of climate change propaganda is based on unproven assertions that have to be taken on faith. Faith in a real religion is a good thing, believing in a power greater than ourselves. Taking man-made climate change on faith is a bad thing. It is based on false data and manipulations by people who benefit from selling you the climate narrative.

At the time of this writing, there is not a single peer-reviewed research study that proves that man-made CO_2 is the control knob of the climate, although many begin with that assumption or use the correlation that temperatures have risen along with CO_2. Therefore, CO_2 is causing the

---

[9] https://www.climatedepot.com/2011/09/14/exclusive-nobel-prizewinning-physicist-who-endorsed-obama-dissents-resigns-from-american-physical-society-over-groups-promotion-of-manmade-global-warming/

temperature to rise. There are many things in life like the sales of ice cream increasing in the summer heat; but ice cream sales didn't cause the summer heat. There are many, many manipulated facts like this that do not in fact support the climate change narrative.

If CO_2 causes temperature to rise, why does it lag the temperature in long term ice cores? If their theory is correct, CO_2 should lead the temperature rather than follow it by about 800 years. How did we have the Little Ice Age, Medieval and Roman Warm Periods, and the Holocene Climate Optimum with relatively stable CO_2? These and many more facts will be presented in more detail later.

Believing in man-made catastrophic climate change gives some peoples' lives purpose and meaning; a reason to get up in the morning, a way to feel good about themselves. As they spit out pieces of their melted paper straw at the traffic light in their EV, they can feel self-righteous, or even better than other people; never giving a thought to their EV tires putting far more nano-plastics into the environment than their old gas car, because the EV tires wear out much faster since they are so much heavier.

For so many in politics, it is satisfying to not only to have a belief, but to have an enemy. It feels good to fight injustice; it is the hero's journey. It is fulfilling to save the world. In the case of climate, it is a falsely righteous path based on made up and manipulated data.

THE DOCTRINE OF THE CULT OF CLIMATE CHANGE

The cult of climate change is similar to a religion in that it provides a moral framework for righteousness (reducing CO_2 emissions and ending fossil fuels) through sacrifice (reversing the prosperity of the Industrial Age) in exchange for salvation (avoiding mass extinction). And they know that endless hardship doesn't sell, so they make up false tales of cheaper energy, and less bad weather, if you just go along with their radical agenda.

It's been said that devoted Christians are less likely to believe in climate change, which I think stems from their faith in God. Christians understand that God wants them to be good stewards of the Earth, and He is in control, not some bureaucrat, politician, or climate cultist. They believe that God blessed the Earth and that it is a gift for man's benefit. They don't believe in overpopulation. They believe that we should go forth and multiply, as God said.

The cult of climate change, however, has an anti-prosperity and even anti-human model. But, of course, they lie to you about it because they know that you would be far less likely to support it if it meant radically changing everything about your life. Many people have no idea the sacrifice that will eventually be required of them and their offspring, on the altar of climate change.

They do not see that instead of leading their families to the Promised Land, they are leading future generations back into slavery and hardship. For the ignorant and misguided, Lord forgive them for they know what they do. For those who know and are in on this anti-prosperity, anti-human, evil agenda, the fires of Hell are just about right. I believe starving people and purposely taking away their prosperity is evil.

The cult of climate change is a bully, like a fear-mongering fire-and-brimstone preacher who focuses less on God's love and more on the message of your sin and damnation to Hell. Cross them, and they will attack you. This cult has caused hysteria and depression for millions of young people since the inception of their propaganda and indoctrination programs.

Grade school kids fear what will happen to them when the Earth overheats. Many fear the future they will face on a too hot Earth. Our civilization and children need saving from climate change energy policies, not from climate change.

BELIEFS & COMMANDMENTS OF THE CULT OF CLIMATE CHANGE

If you want salvation from the coming climate change apocalypse or a second chance at life on a green Earth, you must follow the commandments of the climate change cult. As you saw in an earlier chapter, anybody skeptical about the climate change narrative poses a threat to believers and is labeled a "climate change denier" (aka a realist). Similar to other traumatizing events riddled with mind-control messages in recent years, climate change deniers are shunned and ridiculed for refusing to "follow the science."

In some companies, it will get you fired, demoted, or put out to pasture, never to see a promotion again[10]. According to their religion, deniers should be shunned. If you can make deniers' lives miserable, so much the better. If you can make sure they never get a promotion or their work never sees the light of day, it is good.

In his documentary *Battlefield Melbourne*, Australian filmmaker Topher Field describes how Melbournians were pepper sprayed, tear gassed, shot with rubber bullets, and even imprisoned for "inciting violence," sometimes for something as benign as a social media post questioning the egregious lockdown measures imposed by Premier Daniel Andrews during Covid-19. Naturally, people not complying with lockdown and masking requirements, testing, and, eventually, the vaccines, risked unemployment.

It is my concern that climate change realists will, too, begin to experience backlash from society beyond the obvious social isolation and berating that make their way into headline news. Will climate realists, I wonder, become the new "anti-vaxxers?" Will climate dissidents face jail time? Will they lose the ability to open bank accounts or hold mortgages? Will they be deemed "unhirable?"

Some members of the climate cult are already advocating this and other extreme measures. All is justifiable when one is "saving the world." Sadly, all of these things don't just exist inside of dystopian novels with Orwellian futuristic themes; they are actual concrete realities in today's world.

Because of the migration toward AI, social credit scores, central bank digital currencies (CBDC), universal basic income (UBI), and net-zero initiatives, Covid-19 era lockdowns and measures were probably child's play compared to what's coming down the pipeline…if we don't change course. When we continue to surrender our currencies, banking system, and privacy to the government, we make ourselves vulnerable to being nothing more

---

[10] https://portside.org/2023-01-12/im-scientist-who-spoke-about-climate-change-my-employer-fired-me

than slaves. Because people think they're saving the planet, they'll beg for their invisible chains and be happy to put them on others.

In my own family, I've been shunned by some relatives for my "climate change denial." During Covid it was "you're killing grandma," but today it's more akin to, "you're killing polar bears." When confronted with facts, people get angry and call you names. It is okay for them to have their strongly held opinions taken on faith; it is not okay for you to challenge them with facts that disagree with their feelings.

I learned long ago that there is no point in arguing with leftists and true climate believers. They have their faith and expect you to join them. I advocate that you only engage with those with open minds that can hear the discussion. Because otherwise, you will present facts and they will tell you their feelings and faith…then get angry.

With this level of manipulation, my concern is that, in the end, when armed with climate ideology, the ends will justify the means. I fear that climate realists and anyone who doesn't embrace the government narrative on everything will one day be censored and punished, like totalitarian China today.

If this happens, someday dissenters will be cut off from their food and money supply. And eventually, they may face prison or even death. This is the slippery slope that isn't some melodramatic conspiracy theory; it's a well-documented slide that has been part of communist and totalitarian regimes throughout history. Now they have the technology to do it on a grander, more controlling scale.

Among key phrases used by totalitarian regimes in history is the phrase, "the ends justify the means." This means it is ok to do evil things for their supposed good cause; for some distant Utopia. History is filled with examples of this; some will be provided later in this book.

You are urged to resist the strong-arm tactics of the climate change cult. Instead, become a climate realist, because CO_2 emitting fossil fuels have been the greatest tool of mankind for regular people's prosperity, ever. Evil, whether intentional or not, must be resisted. The climate cult is working diligently to reverse our prosperity.

CLIMATE CHANGE CULT DOCTRINE

To be a member, you must believe that:

1. The Earth is dramatically heating up and it has never been this warm before. (Not true.)

2. CO_2 is the control knob of the climate; and there are no other noteworthy factors in climate. (Not true.)

3. Man-made CO_2 from fossil fuels that we use for energy are the culprits for these ostensibly detrimental temperature increases. Fertilizers, rice, beef, coffee, and other elements of our food supply are also culprits. (Not true.)

4. All non-CO_2-related causes of warming are false, misinformation, and should be ignored, censored, and shouted down. (Not true.)

5. The science is settled. CO_2 is the only source of warming. After all, 97% of scientists agree. (Not true.)

6. Future warming is an existential threat. Climate change will destroy our way of life and beget rising rivers, floods, droughts, hurricanes, and weather disasters of all types, everywhere. (Not true.)

7. The only way to solve climate change is to cut CO_2 emissions drastically by adopting wind and solar or other renewable energies. Wind towers and solar panels are more important than animal habitats, whales, forests, or crop lands. (Not true.)

8. Fossil fuels including coal, oil, and natural gas are bad. Harvesting and using them must stop. (Bad idea.)

9. It is OK for the world's elite to use fossil fuels to power private jets and other luxuries, but the other 99% of the population should be shamed and bullied into zero-carbon lifestyles. (Hypocrisy.)

10. It is OK to ignore the increasing CO_2 emissions of communist China and India provided they continue to fuel our materialistic culture, especially when billionaires and large corporations are the recipients of such revenue. (They become stronger, while the west becomes weaker.)

11. Bonus: It is OK to deny the billions of people who often live in the dark and rely on wood, dung, and crop waste for cooking and heating from using low-cost and readily available coal, oil, and natural gas instead. (Climate imperialism and weak foreign policy.)

It's reasonable for you to be surprised to learn that for the 2023 UN Climate Change Conference (called "COP 28") 100,000 people and all of their food flew to Dubai for this two-week annual climate junket. They ate well and enjoyed air-conditioned venues in this desert location, where the heat is often unbearable. Do they really believe emissions are the culprit, or is it an excuse to party in an exotic location on someone else's cash?

It's also reasonable for you to gawk at the carbon emissions at the 50th WEF annual conference in Davos, Switzerland, where 2,700 corporate millionaires, billionaires, government leaders, and their cronies attended the annual conference to talk about *your* bad behavior and carbon emissions! All of these meetings and travel to get there were made possible by fossil fuels. Needless to say, hypocrisy and self-importance are hallmarks of the climate cult.

Thankfully, people like you are starting to see that the cult of climate change is riddled with false prophets and demands for your sacrifice, not theirs.

CHAPTER FOUR:
BECOMING A HERETIC

> *"Duped Into Supporting IPCC – If the IPCC is dogma,
> then count me in as a heretic."*
> Judith A. Curry, American Climatologist

I am a climate change heretic; and I'm in great company. Alongside me are the courageous Nobel Prize-winning Physicist Dr. Ivar Giaevar, Dr. Judith A. Curry, an American climatologist and former chair of the School of Earth and Atmospheric Sciences at the Georgia Institute of Technology, and many others. Curry was also a member of the National Research Council's Climate Research Committee with over a hundred scientific papers and several major co-edited works to her credit.

Curry retired in 2017 at age 63, coinciding with her public climate change skepticism.[11] During her career, she studied hurricanes, remote sensing, atmospheric modeling, polar climates, air-sea interactions, climate models, and the use of unmanned aerial vehicles for atmospheric research. In short, she is a well-qualified expert on the topic of climate, with a well-informed opinion.

Another brave expert who has spoken out against the climate change hoax is Ronnie Walter Cunningham, an astronaut, fighter pilot, physicist, entrepreneur, venture capitalist, and former employee of NASA as the "third civilian astronaut." In an interview titled "A Conversation with Apollo Astronaut Walter Cunningham about a Vital Need to Restore Climate Science Integrity," Cunningham states that he fears people are unable to distinguish between science and nonscience, "leaving them vulnerable to

---
[11] https://www.epw.senate.gov/public/_cache/files/8/3/83947f5d-d84a-4a84-ad5d-6e2d71db52d9/01AFD79733D77F24A71FEF9DAFCCB056.senateminorityreport2.pdf

the emotional appeal of human caused global warming. Unfortunately, most students today are fed a lot more hype about self-esteem and global warming than real information about history and science. Let's finally recognize that 'self-esteem' is no substitute.[12]"

Until his death in 2023, Cunningham actively wrote and spoke out against what he considered to be the climate change "*hoax*," the idea that humans are affecting the Earth's temperature.

I appreciate Cunningham's willingness to call out the indoctrination of our children with regards to the cult of climate change. One-sided climate change indoctrination is today, an integral part of the public-school curriculum. Teachers who genuinely care about children, and whom the children trust, teach them the unfounded and unproven urgency of catastrophic, man-made climate change. These young minds are exposed to Greta Thunberg, who encourages children to be terrified and threatens pending doom for humankind. They are often shown Al Gore's movie, which is outlawed by court order in the United Kingdom, because of the factual errors and misrepresentations.

Young minds are impressionable. Beating children over their heads with the climate change narrative is at least as abusive to their brains as when we used paddles to punish them in schools decades ago. Many young people are worried sick and depressed about this false teaching and contemplate suicide or giving up hope of a better future.

I have yet to hear of a single elementary, middle school, high school, or college student who has been presented with both sides of the climate change debate. Rather, they are inundated with the "alarmist" side of the argument. At best, they're bullied into accepting the opposing view by a hostile curriculum or teacher. At worst, they're left with trauma.

When I was a young child, I was told that the North Atlantic seabirds, puffins, were endangered. I worked to raise money to save the birds only to later find out that they weren't ever endangered in the first place. The campaign was used to alarm my peers and me to get us emotionally invested in a false narrative.

Major tech platforms actively create barriers to dissenting realist information. Dissenters face removal and censorship, making it more difficult

---

[12] https://www.forbes.com/sites/larrybell/2013/08/06/a-conversation-with-apollo-astronaut-walter-cunningham-about-a-vital-need-to-restore-climate-science-integrity/?sh=3c99eb033345

for non-discerning children and adults to get to the truth. With a major media or university "stamp" of approval, the information is taken at face value, even when it derives from falsified studies, unreliable sources, or is just made up.

There are many wise parents who send their children to reference educational resources such as PragerU, especially in places like California, where the one-sided climate alarmist propaganda isn't just taught—it's the law. Florida, however, drew attention for allowing PragerU videos in their classrooms, providing diverse perspectives on climate change[13]. The CO_2 Coalition also has a great kids' video series.

On March 3, 2024, I simply typed the search term "Florida PragerU climate change" into the Brave browser, and here are my results:

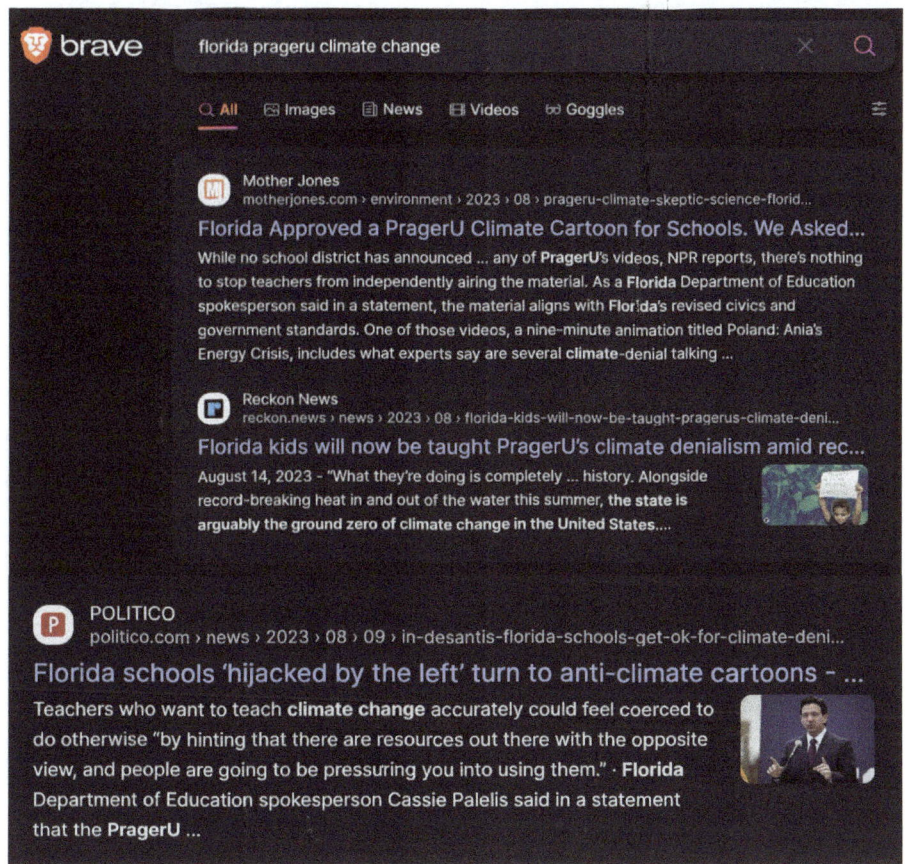

---

[13] https://www.nytimes.com/2023/09/08/us/politics/desantis-florida-storm-climate-change.html

Florida adds PragerU climate denial videos to approved syllabus for children up to 11
irishexaminer.com › homepage

Florida adds PragerU climate denial videos to approved syllabus for...
August 10, 2023 - Despite its name, Prager University Foundation is a lobby group rather than an academic institution. Content on its **PragerU** Kids platform, some of which rejects accepted science about **climate change**, can now be taught to children aged up to 11 in **Florida** public schools. Picture: David McNew/Getty

PragerU
prageru.com › video › is-there-really-a-climate-emergency

Is There Really a Climate Emergency? | PragerU
The **climate** is the most complex system on Earth. Is it really possible to project with any precision what it will be like 20, 40, or even 100 years from now?...

the Guardian
theguardian.com › us-news › 2023 › sep › 06 › prageru-climate-change-denier-republi...

US 'university' spreads climate lies and receives millions from right...
September 6, 2023 - **PragerU**'s influence in pushing false narratives about **climate change** and other far-right shibboleths such as airbrushing the brutal reality of American slavery gained ground when the **Florida** board of education in July gave the green light to using its videos and other materials in classrooms, ...

the Guardian
theguardian.com › us-news › 2023 › aug › 10 › florida-ron-desantis-climate-vidoes-schoo...

Videos denying climate science approved by Florida as state curric...
August 10, 2023 - **Florida**, whose governor Ron DeSantis has called **climate change** "leftwing stuff", is the first state to adopt **PragerU** videos, although in several other states textbooks pushed by the fossil fuel industry have included references that either downplay or deny human-caused global heating.

Snopes
snopes.com › homepage › fact check

Did Florida Education Dept. Approve Climate-Denial Content Comp...
August 10, 2023 - The **Florida** state Department of Education approved content that denies or minimizes the reality of anthropogenic **climate change** and compares **climate** science to Nazism as an optional supplement to the state's social studies curriculum. ... On July 24, 2023, Marissa Streit, the CEO of the nonprofit conservative activist group **PragerU**...

It requires pages of scrolling to find the story, and even when searching "news" on the browser, I only found accusations of PragerU spreading mis-

and dis-information to children or articles about the schools who have rejected Prager Us' videos.

When possible, please bookmark the following websites for you and your family to enjoy. Hopefully, if they are not censored, you will be able to find Climate Truth via the Wayback Machine Internet Archives.

1. Truth In Energy and Climate at www.truthinenergyandclimate.com (my website)
2. CO_2 Science at http://www.co2science.org/
3. CO_2 Coalition at https://co2coalition.org/
4. Real Climate Science at https://realclimatescience.com
5. CEI Competitive Enterprise Institute at https://cei.org/
6. Master Resource at https://www.masterresource.org/
7. The Heartland Institute at https://heartland.org/
8. Climate Depot at https://www.climatedepot.com/
9. No Tricks Zone at https://notrickszone.com/
10. Cornwall Alliance at https://cornwallalliance.org/
11. Science and Environmental Policy Project at https://www.sepp.org/
12. Climate Policy at Heritage at https://www.heritage.org/climate
13. CFACT at https://www.cfact.org/
14. Watts Up with That at https://wattsupwiththat.com/

CHAPTER FIVE:
UNVEILING LIES IN THE MEDIA

*"The Party told you to reject the evidence of your eyes and ears.
It was their final, most essential command."*
George Orwell, *1984*

Dr. John Christy is the Alabama State Climatologist at the University of Alabama in Huntsville with a PHD in atmospheric sciences and B.A. in mathematics. He served as a UN IPCC lead author in 2001 for the 3rd assessment report, and contributed to seven other IPCC reports. His primary interests lie in a satellite remote sensing of global climate and global climate change. He is best known, along with Roy Spencer, for the first successful development of a satellite temperature record.

For his work "developing a global, precise record of Earth's temperature from operational polar-orbiting satellites, fundamentally advancing our ability to monitor climate[14]," Christy was awarded the 1996 Special Award from the American Meteorological Society. In 1991, he received the Medal for Exceptional Scientific Achievement from NASA alongside numerous other professional awards throughout his career.

In addition, Christy authored more than fifty studies published in a dozen peer-reviewed scientific organizations, such as the International Journal of Climatology and Global and Planetary Change.

After such a long and distinguished career as a scientist specializing in climate, Christy stated on May 2, 2007 on CNN that, "I was at the table with three Europeans, and we were having lunch. And they were talking

---
[14] https://www.uah.edu/images/colleges/science/cvs/john-christy-cv-2021.pdf

about their role as lead authors. And they were talking about how they were trying to make the report so dramatic that the United States would just have to sign that Kyoto Protocol."

This brave scientist and whistleblower then detailed how he personally witnessed UN scientists attempting to distort climate science for political purposes. The UN and WEF are commonly the source of mainstream media messaging around climate change. The wind and solar industries have trillions of dollars wrapped up in their narrative, which is how they are capable of purchasing media and advertising to further their agenda, similar to how pharmaceutical companies purchase television advertising. Media companies aren't looking for the best companies to advertise equally. They're taking dollars. Whoever has the most dollars get their message in front of the masses.

In order to capture attention, government-funded (and often falsified) climate and weather studies are paired with sensational clickbait headlines. For instance, in 2019 organizations spent a record $2.4 billion promoting global warming ideology. According to CapitalResearch.org, less than $35 million was invested by "climate realist organizations" in that same year[15,16]. Some of the world's largest corporations are members of the WEF, meaning that climate alarmists and members of the cult of climate change hold the vast majority of wealth in our world, and the advertising dollars to manipulate and traumatize the public with climate alarmism.

Claims to the contrary state that the climate realist organizations are fueled by big oil dollars. However, big oil corporations do not fund the climate realist organizations[17] because they benefit from making their product more challenging to harvest, resulting in higher prices due to scarcity.

---

[15] https://capitalresearch.org/article/environmentalists-spent-a-record-2-4-billion-pushing-global-warming-ideology/

[16] https://www.foxnews.com/politics/meet-little-known-group-funded-left-wing-dark-money-shaping-federal-climate-policy

[17] Truth In Energy and Climate at www.truthinenergyandclimate.com; CO_2 Science at http://www.co2science.org/; CO_2 Coalition at https://co2coalition.org/; CEI Competitive Enterprise Institute at https://cei.org/; Master Resource at https://www.masterresource.org/; The Heartland Institute at https://heartland.org/, Climate Depot at https://www.climatedepot.com/, Real Clear Science at https://www.realclearscience.com/; No Tricks Zone at https://mediabiasfactcheck.com/notrickszone/; Cornwall Alliance at https://cornwallalliance.org/; Science and Environmental Policy Project at https://www.sepp.org/; Climate Policy at Heritage at https://www.heritage.org/climate; CFACT at https://www.cfact.org/; Watts Up With That at https://wattsupwiththat.com/; Science and Environmental Policy Project at https://www.sepp.org/

Chapter Five: Unveiling Lies in The Media

Among the many language manipulations of the cult of climate change is to call their initiatives "green." The "zero-carbon" movement industrial complex has allocated tens of millions to major news outlets and national public radio to propagate climate narratives. This funding incentivizes media outlets to infuse "climate emergency" angles into every conceivable story, and even sporting event reports.

For instance, it was said that the Miami Hard Rock Stadium will soon be underwater due to rising oceans caused by climate change. This story is plainly false. Analyzing historical rates of ocean rise in Florida shows that the ocean will take over a thousand years to reach the parking lot at the current rate of increase from the past 140 years of tidal records.

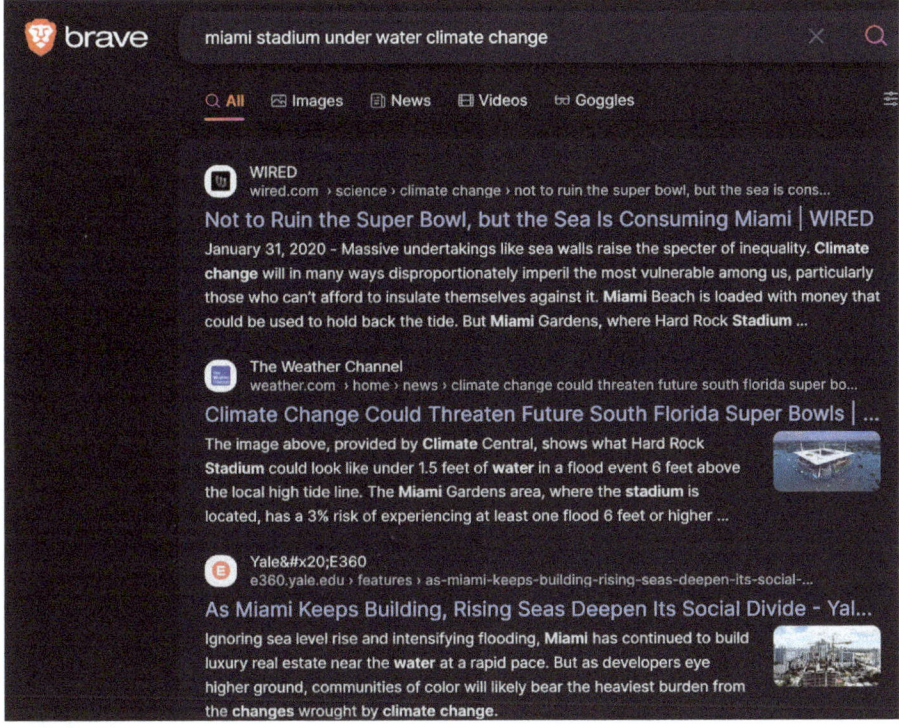

Screen capture from March 3, 2024

THE "97% OF SCIENTISTS AGREE" LIE

I happened to be watching the news when President Barack Obama first said that "ninety-seven percent of scientists agree" that carbon dioxide is causing warming, that it spells doom, and that we must stop it[18,19,20,21]. Although what President Obama stated was a lie, the media ran with it and it was propagandized over and over. Today, that falsehood has morphed into a widespread consensus that man-made climate change is a fact; and "how dare you question or disagree with such common sense?!"

Obama's false reference that "97% of scientists agree," is based on "cooked" evidence designed to claim that there is broad scientific evidence of human-induced climate change[22]. The original paper is authored by Cook, J., Nuccitelli, D., Green, S. A., Richardson, M., Winkler, B., Painting, R., et al., (2013) and it's titled, "Quantifying the Consensus on Anthropogenic Global Warming in the Scientific Literature.[23]"

A review of the actual papers examined by Cook and colleagues revealed that only a small fraction equaling 0.3% of the 11,944 abstracts endorsed the concept of man-made global warming as defined by Cook and his team of research assistants[24]. When taking into account the papers that expressed "no opinion" on man-made causes of climate change, the 0.3% doesn't rise to 97% of scientists agreeing that carbon dioxide causes global warming, but just 1.6% of scientists in this study.

---

[18] https://www.econlib.org/archives/2014/02/david_friedman_14.html
[19] https://wmbriggs.com/public/Legates.etal.2015.pdf
[20] https://www.forbes.com/sites/uhenergy/2016/12/14/fact-checking-the-97-consensus-on-anthropogenic-climate-change/?sh=a1bd79d11576
[21] https://www.instituteforenergyresearch.org/climate-change/the-bogus-consensus-argument-on-climate-change/
[22] https://www.forbes.com/sites/alexepstein/2015/01/06/97-of-climate-scientists-agree-is-100-wrong/?sh=4d5cf9143f9f
[23] https://www.researchgate.net/publication/256538724_Quantifying_the_Consensus_on_Anthropogenic_Global_Warming_in_the_Scientific_Literature
[24] https://wmbriggs.com/public/Legates.etal.2015.pdf

0.3% consensus, not 97.1%

"The scientific consensus that human activity is very likely causing **most** of the **current** GW (global warming)" – Cook et al (2013)

11,944 abstracts (1991-2011) reviewed	100%
7,930 were arbitrarily excluded for expressing **no opinion**	66.4%
3,896 were marked as agreeing we cause **some** warming	32.6%
64 were marked as stating we caused most of the warming	0.5%
41 actually stated we caused most warming since 1950	0.3%
0 were marked as endorsing man-made catastrophe	0.0%

Legates et al 2015 (after Monckton)

In their paper entitled "Climate Consensus and 'Misinformation': A Rejoinder to Agnotology, Scientific Consensus, and the Teaching and Learning of Climate Change[25]," published in April 2015, David Legates and his colleagues discovered discrepancies in Cook's methodology. They found that Cook and his assistants had categorized only sixty-four papers, or 0.5% of the total reviewed, explicitly stating that recent warming was predominantly caused by human activity. Nevertheless, Cook et al. claimed to have identified a "97% consensus" among scientists, even though they clearly only took the opinions of sixty-four hand-selected scientists into account. Cook manipulated data to create a misleading narrative and compelling "soundbite" for the sake of media manipulation.

Simply put, the idea that "97% of scientists agree that carbon dioxide produces climate change) is a lie. Why have the majority of scientists who are either unconvinced or undecided about man-made climate change staying

---

[25] https://wmbriggs.com/public/Legates.etal.2015.pdf "Climate Consensus and 'Misinformation': A Rejoinder to Agnotology, Scientific Consensus, and the Teaching and Learning of Climate Change" D. Legates, W. Soon, Christopher Monckton of Brenchley. Published 1 April 2015
Also found at: https://www.semanticscholar.org/paper/Climate-Consensus-and-%E2%80%98Misinformation%E2%80%99%3A-A-Rejoinder-Legates-Soon/0a39be2c03dc733d5b2bc3b1f4a1a7409aecfc7d

silent? If they don't parrot the narrative, they will lose fellowships, income, and any hopes of further career advancement.

There are others who have done this same analysis of Cook's work and have reached similar conclusions. It is manipulated data[26]. This source includes an article that includes ninety-seven rebuttals to the ninety-seven percent claimed by Cook et al[27].

CENSORSHIP

> *"What historians will definitely wonder about in future centuries is how deeply flawed logic, obscured by shrewd and unrelenting propaganda, actually enabled a coalition of powerful special interests to convince nearly everyone in the world that CO_2 from human industry was a dangerous planet-destroying toxin. It will be remembered as the greatest mass delusion in the history of the world – that CO_2, the life of plants, was considered for a time to be a deadly poison."*
> Richard Lindzen, Professor Emeritus, MIT

A cursory study of history shows that censorship has been used to control people for thousands of years. In recent history, think of Stalin or Lenin. In the past decade, you likely recall the rampant censorship of doctors during the Covid-19 event. Throughout history, we see that sometimes people who question a narrative are censored and made to pay for their heresy.

---

[26] http://www.populartechnology.net/2013/05/97-study-falsely-classifies-scientists.html#Update2
[27] https://climatecite.com/97-articles-refuting-the-97-consensus/

Post

Chris Martz ✓
@ChrisMartzWX

There have been constant attacks and attempts by people to try getting me kicked out of my university because I do not toe the mark. I think for myself and I question authority.

Climate cultists thought all skeptics [or in my case, perhaps more appropriately a "lukewarmer"] were old and dying out. They can't stand the thought of someone as young as myself pushing back on their doomer dogma by sharing meteorological statistics and historical records that contradict their catastrophizing rhetoric.

I am viewed as a threat to their house of cards; they're scared.

David Ippolito @DavidIppol15284 · 18h
Replying to @ChrisMartzWX and @GrnConservatism
Glad to meet another young person who can think critically.

My generation was taught to think critically. We were taught to question authority respectfully but question it
and to look at both sides of an issue

1:12 PM · Feb 10, 2024 · **29.8K** Views

Chris Martz is a young weather historian and climatologist (just a few days younger than Greta Thunburg), who blogs about weather and climate at https://chrismartzweather.com/. Martz eloquently warned his audience on the platform X in February of 2024 that,

> *There have been constant attacks and attempts by people to try getting me kicked out of my university because I do not toe the mark. I think for myself and I question authority. Climate cultists thought all skeptics [or in my case, perhaps more appropriately a "lukewarmer"] were old and dying out. They can't stand the thought of someone as young as myself pushing back on their doomer dogma by sharing meteorological statistics and historical records that*

> *contradict their catastrophizing rhetoric. I am viewed as a threat to their house of cards; they're scared.*

This bright young man is being threatened to have his academic career seized for honestly examining the "facts" of climate change. He recognizes that the propaganda machine doesn't like his work, and that they will try to silence him.

Silencing dissenters is not how we do science. That's censorship.

In 2020, we saw doctors who spoke out against alternative measures taken to treat Covid-19 were silenced, including those who refused to administer vaccines or questioned the use of Remdesivir (sometimes referred to as "run-death-is-near"), as in the case of Dr. Peter A. McCullough, Dr. Sherri Tenpenny, Dr. Christiane Northrup, and Dr. Brian Ardis, among thousands of others.

In the United States Constitution's First Amendment, we are supposed to be protected from the government making laws that, "regulate an establishment of religion; prohibit the free exercise of religion; abridge the freedom of speech, the freedom of the press, the freedom of assembly, or the right to petition the government for redress of grievances."

There is a reason that our Founding Fathers added the 1st Amendment to the Constitution; they sought to establish that our government would not make any type of state religion or limit free speech. However, big tech (and big pharma) is undeniably in collusion with the government in a widespread censorship campaign. Dissenters face censorship or even being de-platformed altogether, as was the case with President Donald Trump on the platform x.com, formerly known as Twitter.

In my estimation, we are in danger of Climate Change becoming part of our official "State Religion." President Joe Biden claims to be using the "all of government approach to climate change.[28]" The order formally establishes the White House Office of Domestic Climate Policy – led by the first-ever National Climate Advisor and Deputy National Climate Advisor – which

---

[28] https://www.whitehouse.gov/briefing-room/statements-releases/2021/01/27/fact-sheet-president-biden-takes-executive-actions-to-tackle-the-climate-crisis-at-home-and-abroad-create-jobs-and-restore-scientific-integrity-across-federal-government/

is a central office in the White House that is charged with coordinating and implementing the President's domestic climate agenda.

The order establishes the National Climate Task Force, assembling leaders from 21 Federal agencies and departments to enable a whole-of-government approach to combating the climate crisis. I suspect that his approach includes collusion with big tech to censor free speech and other dissenters.

There are a few critical facts you're going to take away from this book, but among the most important is this: there is not even one peer reviewed study to support man-made CO_2 as the control knob of the Earth's climate. There are CO_2 studies indeed; I've read a number of them. They begin with the assertion that CO_2 is a controlling factor. However, not a single one provides evidence that CO_2 is the control knob of Earth's temperature changes. If you can find one, please pass it along!

One of the many reasons free speech is important is because truth does indeed set us free. When the government works to hide the truth, it's usually in the interest of controlling people's minds. During 9/11, the images of the falling World Trade Towers, and humans jumping out of windows before the towers fell, permeated the news so that the American people would send their precious sons and daughters to fight in an upcoming war without protest. When the Covid-19 shot campaign began, you heard every radio pundit, newscaster, and even Ted Talk speaker reference the propagandist phrase, "safe and effective," which we now know the shot was anything but.

The U.S. government has been involved in mind control experiments *at least* since the Nazi's began their campaigns. We grabbed the torch and continued mind control experiments on the public that rely on repetition ("safe and effective," 97% of scientists agree, etc.) and horrific images (people jumping out of towers, bodies collapsing in Italy and China, et al) to traumatize people into compliance.

These tools were also used by central planning regimes such as in Cuba and Venezuela, Pol Pot, and others. They sought to control the narrative and silence opposition. For most societies, death follows. We have to learn from history!

When a society surrenders their freedom of speech and critical thinking, they are well on their way to destruction and demise. My own children don't enjoy hearing facts that challenge their rose-colored misconceptions of the world. They'll even go as far as to avoid my wife Amy and me after we gently attempt to inform them of an alternative perspective. In a world of participation trophies, unmerited and unearned self-esteem, and Critical Race Theory (CRT) being taught in public schools, facts no longer invigorate young minds…they hurt their feelings.

Rulers in the above-mentioned totalitarian societies have historically murdered millions of people through central planning. This happened with Lysenko in communist Russia and it starved millions of people in communist China as well.

The improvement and advancement of society, technology, philosophy, and education all require dissenting voices who question the narrative or evidence and propose a better way to think freely, and speak up. The road to hell is paved with good intentions made via consensus or the ruler's desire to control us for our own good.

Case in point, the theory of relativity of how things work, $E=MC2$, was once censored by the Nazis who didn't like Einstein because he was a Jew and was getting attention for his work on the new theory of relativity. They even wrote a book called *100 Authors Against Einstein*, although in the end, they only had 20 authors involved with the project. When Einstein was asked what he thought about a hundred authors being against him and his theory of relativity, it's said that he replied, "Why 100? If I were wrong, it would only take one." This is the basis of science.

Energy equals mass times speed squared ($E=MC2$) was a revolutionary scientific theory in 1905 that's accepted today, but is still referred to as a theory. I wonder why climate change isn't referred to as a theory, too! That should give any intelligent mind pause.

The only way you can "trust the science" is when the science has been scrupulously examined and tested in the real world, standing up against opposition. That is not how "climate science" works today, though.

Here are a few examples of censorship in science. Consensus was defended in these instances, and human progress was hindered.

GALILEO GALILEI

Galileo Galilei, a prominent figure in the 1600s, was an Italian scientist. He faced opposition from the Catholic Church when he asserted that the sun was at the center of the solar system, contrary to the Church's beliefs that everyone believed. Both the Catholic Church and the emerging Protestant Church held a lot of control in Europe. They had the authority of life and death. They burned a lot of people at the stake.

The prevailing belief, supported by religious doctrine, held that mankind was created by God and occupied a central position in the universe. It was widely accepted that the Earth stood at the center of all celestial bodies, and any dissent from this view was heresy. Those who challenged this belief risked being branded a heretic.

Galileo was charged with heresy, which carried the penalty of death. Ultimately, he was found guilty, and his writings were banned. Due to his stature as one of the eras' preeminent scientists, the Church showed leniency in his case.

Despite his significant contributions to science and thought, Galileo spent the rest of his life under house arrest. It wasn't until nearly three centuries later that the Catholic Church acknowledged its error and officially absolved Galileo of heresy, recognizing the validity of his scientific findings.

Galileo's experiment at the Leaning Tower of Pisa famously disproved Aristotle's long standing scientific consensus that heavy objects fall faster than lighter ones. By dropping two cannonballs of equal size but different weights from the tower, Galileo demonstrated that they fell at the same speed.

His experiment revealed that weight did not influence the rate of descent for objects of the same size. This groundbreaking finding challenged the consensus of the time and further solidified Galileo's reputation as a pioneer in scientific inquiry.

TROFIM LYSENKO

Trofim Lysenko, a prominent Soviet biologist, wielded immense influence, promoting groupthink and enforcing the Soviet consensus. This resulted in catastrophic widespread starvation. Despite his mistaken beliefs regarding crop cultivation for increased yields, Lysenko possessed the authority to censor and eliminate dissenters from his pseudo-scientific theories.

His misguided ideas, endorsed by the totalitarian regime, controlled all facets of life; stifling individual expression; and imposing state-sanctioned beliefs. This suppression paralleled modern-day scenarios observed in communist China, Islamist Iran, Russia under dictatorship, as well as socialist regimes like Cuba and Venezuela. It is creeping into our lives here in the United States and the Western world as well, in the climate change arena.

Under Lysenko's influence, genetic research was banned in the Soviet Union, hampering agricultural advancements, and hindering scientific progress. Farmers were coerced into adopting inefficient practices based on Lysenko's flawed principles, causing crop failures and starvation.

The regime, led by Stalin, enforced compliance with Lysenko's doctrines, silencing dissent through oppression and imprisonment. Scientists who refused to abandon genetics faced severe consequences, losing their livelihoods and freedoms. Many endured imprisonment, like Nikolai Vavilov, who was sent to a concentration camp.

Lysenko's policies caused famines during which millions of people starved to death in the Soviet Union. The adoption of his methods by the People's Republic of China in 1958 led to the devastating Great Chinese Famine from 1959 to 1962. Communist solidarity prompted China's adherence to failing policies, despite their disastrous consequences.

The resurgence of cooperation between China and Russia poses a significant risk, reminiscent of past alliances founded on dangerous ideologies. It underscores the enduring threat posed by totalitarian regimes and the potential for catastrophic outcomes when flawed policies are enforced without dissent.

ANTÓNIO EGAS MONIZ

In 1949, Portuguese neurologist António Egas Moniz was awarded the Nobel Prize in Medicine for his role in popularizing and inventing lobotomies. This procedure involved drilling a hole in the forehead of a mentally ill person and rooting around in the brain, causing significant brain damage.

Initially hailed as a breakthrough treatment, lobotomies were embraced as consensus medical science. However, over time, the detrimental effects and ethical concerns from the procedure became apparent, leading to its eventual abandonment. Lobotomies serve as an example of how medical practices, once deemed effective, can later be recognized as harmful and outdated.

COCAINE, THE WONDER DRUG?

During the Victorian era, cocaine was hailed as a "wonder drug" with widespread popularity, and the prevailing consensus was overwhelmingly positive about its benefits.

However, as time progressed, the strongly addictive nature of cocaine use became evident. The initial consensus regarding its benefits eventually gave way to an understanding of its harmful effects. This shift highlights how societal perceptions and consensus on substances like cocaine can evolve over time as more information becomes available about their consequences.

Promoting scientific integrity, transparency, and open discussion is crucial for understanding our world. Encouraging open debates allows different viewpoints to be heard and the challenging of established consensus ideas. Real science thrives on skepticism and the ability to question the status quo.

In the realm of climate science, however, debates are rare. Genuine science should encourage other perspectives and a rigorous examination of evidence. Presenting only one side of the argument without allowing dissenting opinions or beliefs is propaganda, not scientific discourse.

Part Two:
CO_2 LIES

CHAPTER SIX:
THE CO₂ LIE

> *"We have no reason to think that climate change is harmful if you look at the world as a whole. Most places, in fact, are better off being warmer than being colder. And historically, the really bad times for the environment and for people have been the cold periods rather than the warm periods."*
> Dr. Freeman John Dyson[29]

CO_2 lies are the foundation of most fear mongering and economic initiatives around climate change, so understanding the basics about CO_2 propaganda is important to help you prepare for the remaining climate lies we'll discuss throughout this book.

If you're unsure about my larger premise of climate lies being weaponized as a tool of control, this chapter will help to open your eyes.

The push for decarbonization, especially by 2050, is absurd. The CO_2-less "energy transition" with its emphasis on wind, solar, and electric vehicles isn't just unnecessary, it is expensive. Contrary to what you have been told, wind and solar aren't cheaper; if they were, your electric bill would be going down, not up.

Calling CO_2 "carbon" is like calling water "hydrogen" …both are just plain false. Calling CO_2 carbon is a matter of marketing. It's designed to make you think of charcoal and soot, not your breath and clean air. You see,

---

[29] **Dr. Freeman John Dyson**, a British-American theoretical physicist and mathematician, left an indelible mark on various scientific fields. Renowned for his contributions to quantum field theory, astrophysics, random matrices, mathematical formulation of quantum mechanics, condensed matter physics, nuclear physics, and engineering, his intellectual legacy is extensive. His distinguished career included serving as a professor emeritus at the Institute for Advanced Study in Princeton. Additionally, he played a crucial role as a member of the board of sponsors of the Bulletin of the Atomic Scientists.

all humans exhale about seven hundred pounds of CO_2 per year. Throughout this book, I'll share different ways that our "carbon emissions" actually *benefit* the environment. Trees and plants consume our CO_2, and then they provide us with oxygen.

The carbon lie is that if we pump more CO_2 into the atmosphere, we will catastrophically alter our environment. The amount of CO_2 you or your business emits is also commonly referred to as your "carbon footprint." However, without carbon dioxide in the Earth's atmosphere, there would be no life on Earth. CO_2 isn't a pollutant; it's essential for life to exist!

The Intergovernmental Panel on Climate Change (IPCC), whom we'll reference throughout this book, posits that 89% of our carbon emissions are derived from fossil fuels and industry. Their initiative includes reducing those emissions to zero by the year 2050 if we want to save humanity from destruction. This is an impossible task.

Understanding that all life is carbon-based is important. All life originates from carbon dioxide and oxygen in the atmosphere. All the CO_2 we "emit" from fossil fuels today was once CO_2 in the atmosphere. Coal, oil and natural gas are the products of plants and plankton that transformed into fossil fuels over millions of years.

In fact, historically carbon dioxide in the atmosphere has been much higher than it is today. CO_2 levels were as much as 10 times higher than they are today when life on Earth began, often referred to as the "Cambrian Explosion" of life. The ideal condition for plant growth is in an atmosphere that has four to five times the amount of CO_2 that is in our atmosphere today.

Most importantly, higher CO_2 levels will boost forest and farming productivity, which is beneficial for our population of eight billion people that may increase to as many as ten billion souls by 2050. "Satellite measurements have noted the 'greening of the Earth' as crops and forests grow better due to our higher levels of CO_2," as reported by NASA.

If we really care about the whole of humanity, we will not eliminate CO_2 emissions, but celebrate the greener Earth they are causing. When the Earth is greener, food costs decline and wildlife thrives. However, when energy prices skyrocket, food costs skyrocket too. In short, the real social justice model is one that encourages CO_2 emissions and keeps food and energy prices down.

The "energy transition" is a threat to humankind, full stop. The organizations I mentioned, like the United Nations, the World Economic Forum, and leftist politicians will continue selling climate change propaganda until we put our collective foot down to stop their madness.

They don't hate pollution. They hate the humans releasing CO_2 and the "useless eaters" they see us as. We do not need saving from climate change. We need saving from these corrupt globalists, particularly at the WEF and UN.

I've been called a climate heretic for my commitment to going against the church of climate change. I hope you'll defect with me so we can lead humanity toward a greener, cleaner, and brighter future.

WHAT IS "THE GREENHOUSE EFFECT?"

> *"All of the models that predict catastrophic global warming fail the key test of the scientific method: they grossly overpredict the warming versus actual data. The scientific method proves there is no risk that fossil fuels and carbon dioxide will cause catastrophic warming and extreme weather."*[30]
> William Happer, Ph.D

Carbon dioxide serves as a vital nutrient for plants, enhancing growth rates, increasing heat tolerance, and water efficiency—a phenomenon exploited in greenhouse agriculture. Greenhouses routinely add three to five times as much CO_2 as the air naturally contains. Mainstream media overlooks the benefits of increased CO_2 levels, and instead focuses on alarmist propaganda.

Dr. William Happer, Ph.D., a Professor Emeritus of Physics at Princeton University, specializes in modern optics; that is optical and radio frequency spectroscopy of atoms and molecules, radiation propagation in the atmosphere, and spin-polarized atoms and nuclei. He boasts a substantial

---

[30] https://co2coalition.org/news/two-princeton-mit-scientists-say-epa-climate-regulations-based-on-a-hoax/

academic background, having authored over 200 peer-reviewed scientific papers.

Beyond academia, Dr. Happer has held key roles such as the Director of Energy Research in the U.S. Department of Energy. Additionally, he has chaired the Steering Committee of JASON, a group of scientists and engineers providing counsel to Federal Government agencies on defense, intelligence, energy policy, and other technical challenges.

I had the pleasure of sitting down with both Dr. Happer and Dr. Richard Lindzen for a meal during which I found them both to be well-informed, professional, and remarkably pleasant. The following comes from a Wayback Machine, originally published by the *Epoch Times*[31]:

> *Citing extensive data to support their case, William Happer, professor emeritus in physics at Princeton University, and Richard Lindzen, professor emeritus of atmospheric science at Massachusetts Institute of Technology (MIT), argued that the claims used by the EPA to justify the new regulations are not based on scientific facts but rather political opinions and speculative models that have consistently proven to be wrong.*
>
> *'The unscientific method of analysis, relying on consensus, peer review, government opinion, models that do not work, cherry-picking data and omitting voluminous contradictory data, is commonly employed in these studies and by the EPA in the Proposed Rule,' Mr. Happer and Mr. Lindzen stated. 'None of the studies provides scientific knowledge, and thus none provides any scientific support for the Proposed Rule.'*
>
> *'All of the models that predict catastrophic global warming fail the key test of the scientific method: they grossly overpredict the warming versus actual data,' they stated. "The scientific method proves there is no risk that fossil fuels and carbon dioxide will cause catastrophic warming and extreme weather.'*
>
> *'Climate models like the ones that the EPA is using have been consistently wrong for decades in predicting actual outcomes,' Mr. Happer told The Epoch*

---

[31] https://web.archive.org/web/20230830100556/https://www.theepochtimes.com/article/two-princeton-mit-scientists-say-epa-climate-regulations-based-on-a-hoax-5460699

> Times. To illustrate his point, he presented the EPA with a table showing the difference between those models' predictions and the observed data.'

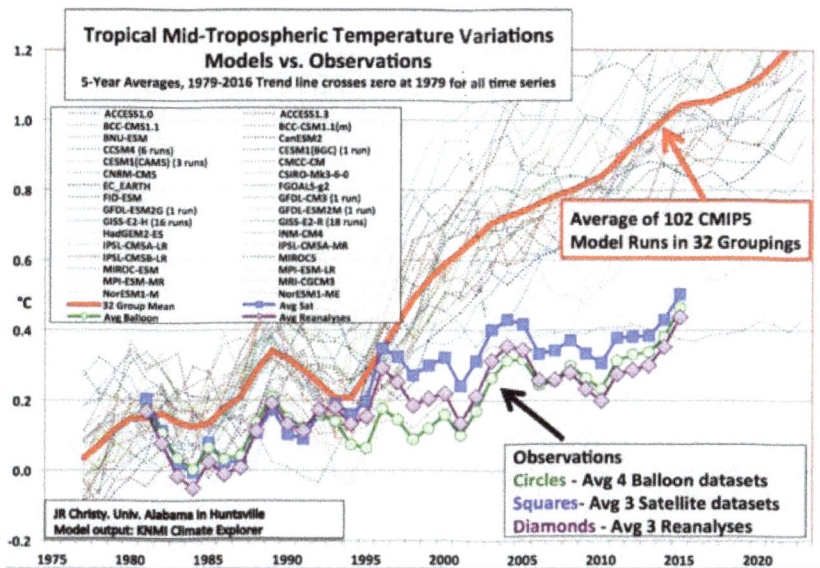

Modeled climate predictions (average shown by red line) versus actual observations (source: J.R. Christy, Univ. of Alabama; KNMI Climate Explorer)

The Greenhouse Effect is the concept that the Earth is warming and it is mainly caused by CO_2, methane, and nitrous oxide. In truth, the Greenhouse Effect keeps Earth's temperature livable. In the simple graphic that follows courtesy of the CO_2 Coalition, a source I trust with scientifically sound information, we read that "about 70% of sunlight is absorbed by the surface of the Earth and converted to heat. The warm surface, greenhouse gases and clouds convert the heat to long wave infrared radiation (LWIR) Some of the LWIR escapes to space, but some radiated back down to Earth's surface, which makes it warmer than it would be with no greenhouse gases[32]."

---

[32] https://CO2coalition.org/

All greenhouse gases do pretty much the same thing: they absorb and reradiate heat in all directions. Heat radiating from the Earth, which is warmed by the sun. Most people don't know that water vapor is dominant and most abundant. The science community leaves it off of the charts of what is in Earth's atmosphere; the climate alarmists never talk about it.

Chapter Six: The CO₂ Lie   61

Water vapor varies from about 0.25% at the south pole, where it is extremely cold almost all the time, to about 4% at the equator where it is often in the eighties, in degrees Fahrenheit. Water vapor is the most important abundant greenhouse gas. It is fifty times more abundant than CO_2, 10,000 times more abundant than methane, or natural gas. It is 60,000 times more abundant than nitrous oxide.

Graph created by Frank Lasee

In the graph above, there are 10,000 dots, illustrating the ratios of each molecule in the atmosphere. Water vapor is regularly left out of graphs like this. However, water vapor is an important component of the atmosphere and should not be ignored. My suspicion is that eliminating water vapor is part of the climate change cult propaganda, because water vapor, not CO_2, is the most important, abundant greenhouse gas.

GREENHOUSE GASES MATTER

Greenhouse gases do matter. Water vapor is the most abundant greenhouse gas, at about 20,000 parts per million (ppm) or 2%, CO_2, at 420 ppm or .042%., methane at 1.8 ppm or .00018%, nitrous oxide is about 0.3 ppm or 0.00003%.

According to the EPA, a part per million is about one drop of liquid diluted into 13 gallons of liquid. This means that the air is like 13 gallons of liquid having 20,000 drops of water, 420 drops of CO_2, not quite 2 drops of methane, and a third of a drop of nitrous oxide. And the climate alarmists are focusing on the 422.3 drops rather than the 20,000 drops.

At present rates of increase it will take 200 years for CO_2 to double to about 840 ppm; 270 years for methane, and about 270 years for nitrous oxide. There is a lot of time before any of these double.

Water vapor, however, will not double. Water vapor is constrained by temperature. When warm moist air meets cold air, water vapor turns into rain, snow or it can condense as dew. It cycles out of the atmosphere into water on the surface, before starting the cycle over again.

Those climate alarmists who know that water vapor is the dominant greenhouse gas say that CO_2 makes everything warmer and a warmer atmosphere will hold more water vapor, causing more warming and creating runaway global warming.

This always bothered me, but I didn't know why. Then it dawned on me...

Just because warmer air *can* hold more water vapor, doesn't mean that it will. Think of the great and small deserts of the world, such as Death Valley, the Sahara, the Gobi Deserts, and the dry southwestern United States. Most of Australia is very dry, too. Do you notice one thing these warm climates have in common? They're also dry. Warmer air isn't always wetter.

But what about doubling CO_2? Won't that have some sort of catastrophic or at least monumental effect on the Earth?

Nope.

Doubling CO_2 to 800 ppm will only have a 1% warming effect; it'll take nearly 200 years to get there. The warming process is very slow.

In a YouTube presentation[33] given by Dr. Happer, the Princeton professor emeritus mentioned at the beginning of this chapter, we learn about how much more CO_2 will warm at 800 ppm, and as the chart below illustrates, it's very little. As the presentation shows, the sun heats the Earth and greenhouse gasses hinder the cooling. More CO_2 is a minor player compared to water vapor.

The following is a graph from Dr. Happer's' calculations showing this concept.

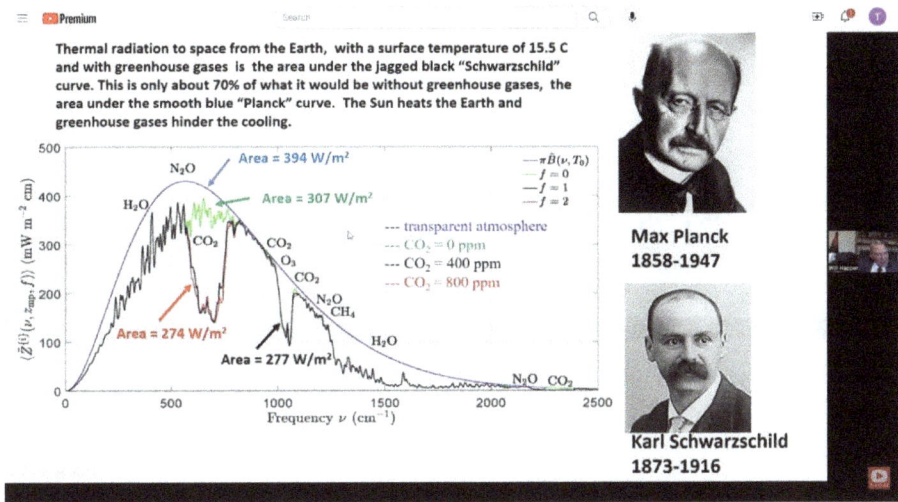

Source: "CO_2, The Gas of Life" Dr. William Happer Physicist Princeton
https://www.youtube.com/watch?v=tXJ7UZjFDHU

---

[33] https://www.youtube.com/watch?v=CA8elCE75ns

I invite you to read Wijngaarden and Happer's 50-page study called "Infrared Forcing by Greenhouse Gases" (link in footnotes[34]).

Some might ask why Happer's work hasn't been peer-reviewed. It is because the climate alarmists will not allow that to happen. Dr. Dick Lindzen of MIT had two of his research studies, which were critical of CO_2 as the sole cause of warming, published in scientific journals in the 1990s, and the editors were summarily fired. This is how climate change science works; by condemning science and anybody who stands in the way of the narrative.

Doubling CO_2 in the atmosphere is like putting a second coat of high-quality red paint on the same barn. It doesn't get much redder than after the first coat.

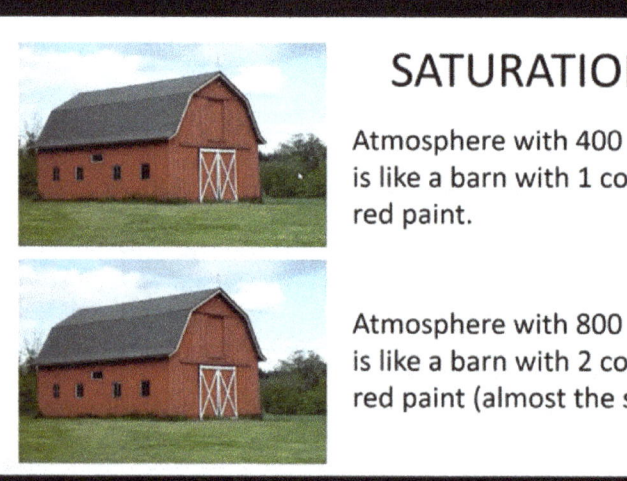

Source: "CO_2, The Gas of Life" Dr. William Happer Physicist Princeton
https://www.youtube.com/watch?v=tXJ7UZjFDHU

Each additional CO_2 molecule warms a little less than the one before it. We learn from Dr. Happer that water vapor and clouds make up 95% of the greenhouse effect, *not* CO_2. In order to learn more about Happer's studies on clouds, please visit the link in the footnotes, where you'll find a variety of sources to examine cloud and greenhouse gas studies.

---

[34] https://CO2coalition.org/wp-content/uploads/2022/03/Infrared-Forcing-by-Greenhouse-Gases-2019-Revised-3-7-2022.pdf

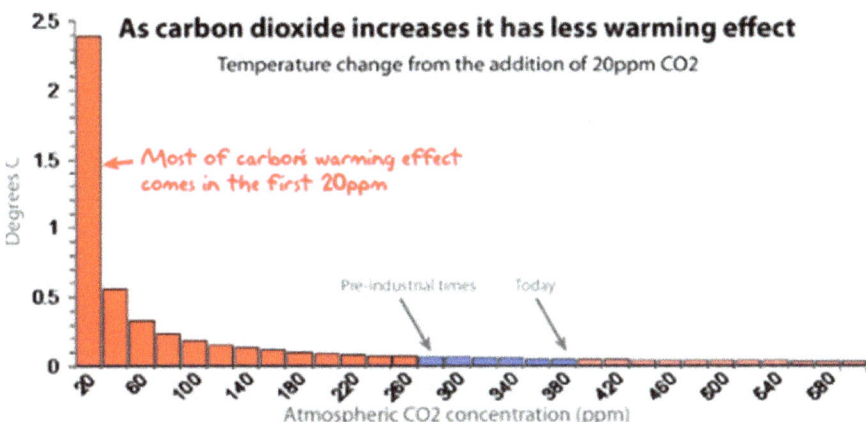

Source: "CO_2, The Gas of Life" Dr. William Happer Physicist Princeton
https://www.youtube.com/watch?v=tXJ7UZjFDHU

CHAPTER SEVEN:
CO₂: THE MIRACLE MOLECULE

> *"We know that carbon dioxide has been a much larger fraction of the Earth's atmosphere than it is today, and the geological record shows that life flourished on land and in the oceans during those times. The incredible list of supposed horrors that increasing carbon dioxide will bring the world is pure belief disguised as science[35]."*
> Harrison Hagan Schmitt

If you've ever used Miracle Grow, you've seen the remarkable impact it has on plant growth: bigger fruits, more vibrant flowers, and enhanced colors. Plants thrive faster and healthier when given Miracle Grow compared to those grown in nutrient-depleted soil.

CO_2 is an airborne miracle growth agent. Most plants thrive at 1,200 to 2,500 ppm (parts per million), which is three to five times more CO_2 than the current 420 ppm. Plants are naturally adapted to higher CO_2 levels. More plant growth is good for all living creatures (including humans) and the Earth itself. Plants and ocean plankton, crucial sources of food, thrive in a CO_2-enriched environment.

At 150 ppm of CO_2, plants cannot survive, which poses a threat to all life on Earth. More CO_2 benefits the oceans, which are the real "lungs of the earth." Phytoplankton make 80% of Earth's oxygen and they flourish with increased CO_2, making the ocean ecosystem healthier and more productive. Sea life thrives when it has more food.

---
[35] https://www.wsj.com/articles/SB10001424127887323528404578452483656067190

Doubling, tripling, or even reaching 2,000 ppm of CO_2 over several centuries will not lead to overheating. The Earth operates as a self-regulating system, with negative feedback mechanisms maintaining equilibrium to stay in balance.

This scientifically-supported perspective contradicts prevailing climate narratives and propaganda. Advocating for more CO_2 is a challenge to mainstream climate beliefs, emphasizing that its positive impact on plant life, and overall environmental health is climate heresy. The cult of climate change doesn't want you to get any impressions that you can find salvation in any church but theirs, and they want you to work out your faith in trembling fear; only compliance will do.

Plants, essential to all life, evolved when CO_2 levels were much higher than today's 420 ppm. And, CO_2 promotes the growth of microbes, roots, and worms all of which improve the soil. The Earth is 35% greener over the last 45 years and the NASA study attributes 70% of that to the increased CO_2[36].

This image shows the change in leaf area across the globe from 1982-2015.
Credits: Boston University/R. Myneni

---

[36] https://www.nasa.gov/technology/carbon-dioxide-fertilization-greening-earth-study-finds/

Flawed studies claim reduced nutrient content in food crops due to increased CO_2 fail to consider the availability of essential nutrients. When nutrients exist in food crops, the plants absorb them. Their roots grow deeper and wider with more CO_2 as the plants seek more nutrients to fuel their growth. These healthy, well-rooted plants improve soil quality, too. More CO_2 means more nutrient-dense foods, which increases harvests, improves soil, and improves our health.

Some people confuse carbon monoxide, which is poisonous, with carbon dioxide (CO_2), wrongly assuming that it, too, is poisonous. Recognizing this distinction is critical, and understanding the role of CO_2 in the circle of life and photosynthesis would end the unnecessary vilification of carbon dioxide.

In conclusion, CO_2 is a "miracle trace gas" that contributes to a greener planet, increased food production, and enhanced soil health. Acknowledging scientific evidence and promoting informed discussions are crucial to dispelling climate change myths.

CO2: THE CATALYST FOR CROP GROWTH

"How humanity and the rest of the biosphere will prosper from this amazing trace gas (CO_2) that so many have wrongfully characterized as a dangerous air pollutant."
Dr. Craig Idso, Ph.D

Dr. Craig Idso, Ph.D., serves as the Founder and current Chairman of the Center for the Study of Carbon Dioxide and Global Change, a non-profit organization committed to exploring and disseminating scientific insights on the impacts of atmospheric carbon dioxide enrichment on climate and the biosphere.

Dr. Idso's research has been featured in numerous peer-reviewed journals, and he has authored or co-authored several books on related subjects. His academic journey includes a B.S. in Geography from Arizona State University, an M.S. in Agronomy from the University of Nebraska – Lincoln, and a Ph.D. in Geography from Arizona State University.

Dr. Idso has contributed significantly to the field, having been a faculty researcher in the Office of Climatology at Arizona State University. Additionally, he has shared his expertise through lectures, leading enriching academic discourse on climate-related topics.

Increasing Levels of Carbon Dioxide (CO2)

In the image above, you'll see four images of Dr. Idso holding cards next to four trees that were given the same nutrients and water over the same period of time. In the first image he is holding a card that says AMB, which represents about 380 ppm of CO_2. The second image shows an identical species tree flourishing from a small shrub to a proper tiny Christmas tree, and that tree flourished with about 530 ppm of CO_2. The third tree grew even more at 680 ppm and finally, in the fourth image, Dr. Idso is standing next to a tree having grown nearly as tall as the scientist at 730 ppm. The experiment shows that CO_2, being the only variable in the experiment, helps the trees flourish.

Dr. Idso states, "Based on the numerous experiments... I can tell you that, typically, a 300-ppm increase in the air's CO_2 content... will raise the productivity of most herbaceous plants by about one-third, which stimulation is generally manifested by an increase in the number of branches and tillers, more and thicker leaves, more extensive root systems, and more flowers and fruit[37]."

My grandmother told me to talk to my house plants; she always had a green thumb. Well, you know what? Grandma was right. Talk to your plants and get close to them; they flourish by absorbing the CO_2 you exhale and will thank you by providing more oxygen in your house. In a single year, a person exhales about seven hundred pounds of CO_2, and provides the planet with the miracle growth it needs to flourish.

In September of 2023, the University of Oklahoma published a study entitled, "Greenhouse Carbon Dioxide Supplementation," which states, "In general, CO_2 supplementation is the process of adding more CO_2 in the greenhouse, which increases photosynthesis in a plant. Although benefits of high CO_2 concentration have been recognized since the early 19th century, growth of the greenhouse industry and indoor gardening since the 1970s has dramatically increased the need for supplemental CO_2[38]."

Plants don't need less CO_2, they need more of it.

DUE TO MORE CO₂ FOOD CROP YIELDS ARE ON THE RISE

At the time of writing this book, food crop harvests and yields are on a continuous upward trajectory globally, even in regions lacking access to advanced seeds and fertilizers, as can be seen in the graph below provided by Our World Data. The graph illustrates the evolution of crop yields worldwide from 1961 to 2020, with wheat, corn, soybeans, and rice standing out as the top four contributors to global food and animal feed. These staple crops have experienced substantial increases in harvest size and yields. This fact alone negates the argument that rising CO_2 is leading us down the road to food scarcity, as the common narrative would have you believe. In the real world,

---

[37] https://www.masterresource.org/carbon-dioxide/increased-plant-productivity-the-first-key-benefit-of-atmospheric-CO2-enrichment/
[38] https://extension.okstate.edu/fact-sheets/greenhouse-carbon-dioxide-supplementation.html

not the upside-down world of the WEF and UN pundits and propaganda media, an increase in CO_2 actually spells abundance, not hunger.

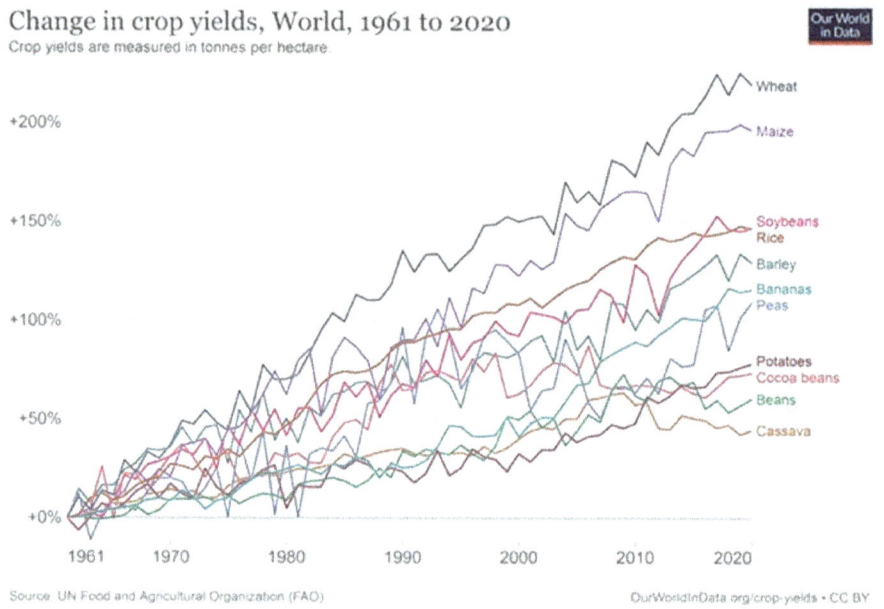

Photo Courtesy of UN Food and Agricultural Organization (FAO)

One important thing to note is that a crop "harvest" represents the total produce available for consumption from a crop, while crop yield measures the amount harvested per unit of land. Climate alarmists tell us just the opposite in their dystopian talking points. This trend is happening in all nations, and across all measured crops as well such as in rice, coffee, wheat, and others.

Crop yields for staples like wheat and corn (maize) have tripled since 1960, a remarkable achievement, quietly ignored by doomsday climate predictions. Today we are feeding an increasing population with less land than ever before. Considering how much land we waste with wind turbines and solar panels that decimate environmental habitats, that's great news! Coffee is increasing, too. Despite media reports suggesting a decline in coffee production, the graph shows a healthy upward trend, with a temporary setback during the Covid period. This isn't at all what the media has reported.

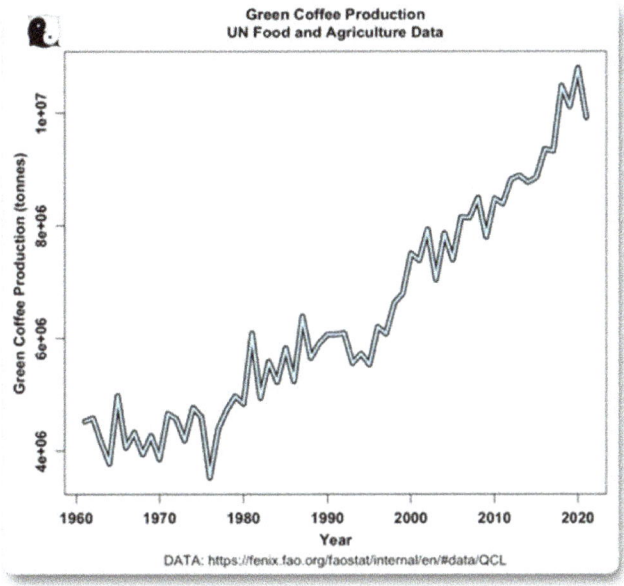

Rice is among the greatest illustrations of crops thriving with increased CO_2. In the graph below, you'll see that rice grows far better with more CO_2. An increase from 420 CO_2 ppm, where we are today, to 800 ppm as is illustrated in this picture shows much healthier, vibrant rice plants.

Source: Susanne von Caemmerer, W. Paul Quick, and Robert T. Furbank (2012). *The Development of C4 Rice: Current Progress and Future Challenges. Science* 336 (6089): 1671-1672 https://notrickszone.com/2012/12/29/higher-co2-concentrations-will-feed-a-billion-more-people/

Banana harvests and yields are also up since 1961 according to Our World Data. Below, you'll see that South America, Africa, and Asia are experiencing five times the banana production in 2020 than they were in 1961, despite potential threats from a banana-killing fungus that still poses a risk to global banana production in the future. Thankfully, with more CO_2 we also have a healthier microbiome for plants. Hopefully that will make our crops less vulnerable to invasive pests or fungi.

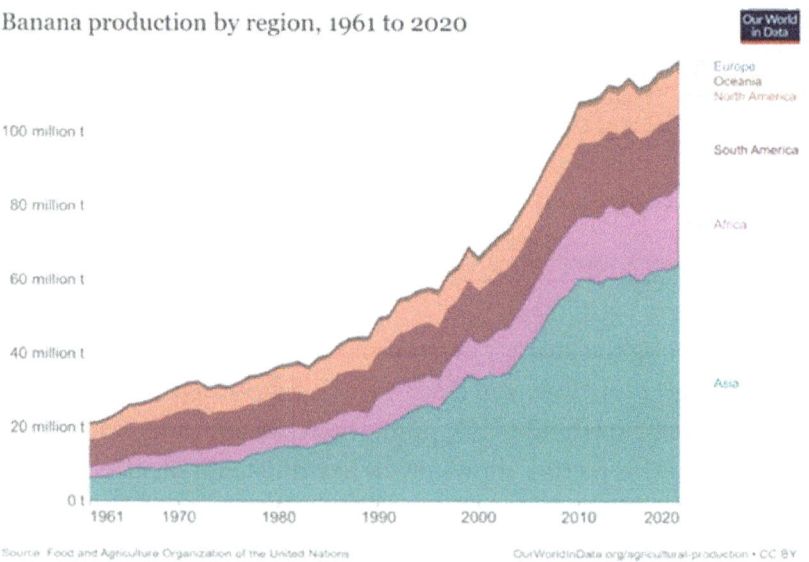

Dr. Craig Idso and Dr. Sherwood Idso emphasize the positive impact of increased CO_2 levels in their book titled *The Many Benefits of Atmospheric CO_2 Enrichment*. The Idso's comprehensive resource delves into numerous studies that validate the many benefits of elevated CO_2. For those interested in exploring this topic further, the book is a valuable reference, providing evidence-based insights into the many positive outcomes from higher CO_2 concentrations.

Some countries are attempting to reduce or eliminate the use of nitrogen fertilizers made with natural gas. Sri Lanka's experiment resulted in a drastic 40% decline in food crop and export yields, inflicting significant hardship on their people[39]. Coincidentally, Sri Lanka had a near perfect ESG score as they decimated crop and export yields to nearly *half* of their former production. A source states, "last year, we got 60 bags from these two acres. But this time it was just 10[40]."

From an article which appeared in *Reuters* on November 11, 2014:

> *"During the main rice cultivation season in 2019, Sri Lanka produced 3.5 billion kg of the grain. Agriculture experts predicted paddy output could fall as much as 43% this year due to the import ban.*
>
> *Jeevika Weerahewa, an agricultural economist at Sri Lanka's Peradeniya University, said paddy yields could drop by 30% and maize by 50% even with the import ban reversed.*
>
> *'This was an unnecessary experiment,' she said. 'The time for fertilizer application for rice is past and the crop will not recover. In addition, world market prices have more than doubled and there are no suppliers.'*
>
> *Weerahewa said Sri Lankas' weak foreign exchange reserves, which dropped to $2.27 billion at the end of October, would impede imports. She said rice yields were not expected to recover until at least the second quarter of 2022[41]."*

The problem persists in Sri Lanka. In May of 2022, Al Jazeera reported:

> *Walsapugala, Sri Lanka – Mahinda Samarawickrema, 49, will not be planting paddy this season.*
>
> *After a government ban on chemical fertilizers cut his rice yield in half during the March harvest, the farmer, who owns eight hectares (20 acres) of paddy and banana, said he no longer has the income to maintain a farm. Especially as his banana crop also looks set to fail.*

---

[39] https://www.reuters.com/markets/commodities/fertiliser-ban-decimates-sri-lankan-crops-government-popularity-ebbs-2022-03-03/

[40] https://www.reuters.com/markets/commodities/fertiliser-ban-decimates-sri-lankan-crops-government-popularity-ebbs-2022-03-03/

[41] https://www.reuters.com/markets/commodities/sri-lanka-rows-back-organic-farming-goal-removes-ban-chemical-fertilisers-2021-11-24/

The cult of climate change got what they wanted. Sri Lanka has a near-perfect ESG score (98) which is higher than Sweden (96) or the United States (51)[42]. They also can't feed their people and have run out of nearly everything.

Sri Lankan leaders mistakenly followed the UN IPCC recommendations of eliminating natural gas made nitrous oxide from nitrogen fertilizers. Half of the world relies on man-made fertilizers for their farming.

Like CO_2, nitrogen is necessary for most food crops. As we discussed earlier in this chapter, nitrous oxide is a tiny part of our atmosphere. Remember it is 60,000 times less abundant than water vapor, the dominant greenhouse gas.

Nitrous oxide is inconsequential to climate change.

Limiting fertilizer isn't a decision made with common sense. Fertilizer is a pawn in the climate change cult's game of achieving more control and power over more people. Stopping the use of man-made fertilizers is unscientific and absurd. Every day that we don't stand up to these lies and stupidity, we are closer to the death of millions, if not billions, of people.

The following graph from the CO_2 Coalition shows how the incredible increase in grain yields track the same as the modest growth in CO_2. CO_2 is plant food, just like man made nitrogen fertilizers.

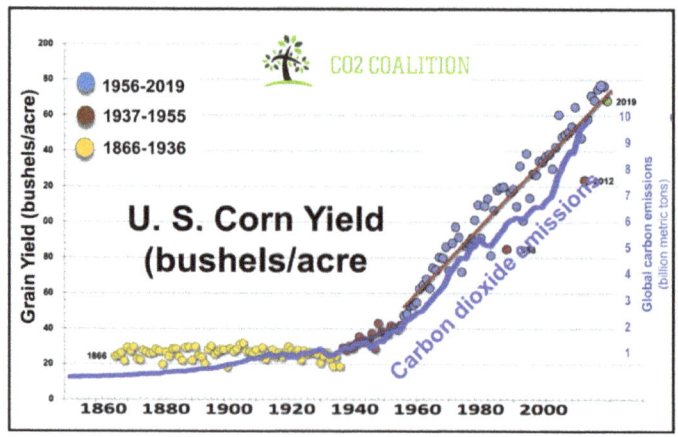

Graph courtesy of the CO2 Coalition[43]

---

[42] #
[43] https://CO2coalition.org/

Among the many alarms that will go off when you read this book, the fact that climate policies decimate food production as well as how much land we have available for agriculture should be setting off some bells. Limiting food production *on purpose* in a world with a rapidly growing population must be stopped; and it is my intention that *Climate and Energy Lies* will help everyone wake up, understand the truth, and do their part to end the crazy climate, energy, and agriculture policies. There is no plan to replace nitrogen fertilizers with an alternative that will allow farmers to continue harvesting what they are currently producing.

Efforts to curtail the use of these fertilizers have been spearheaded in countries such as the Netherlands, Ireland, Canada, and New Zealand. Curtailing nitrogen fertilizers and limiting the use of crop land, as these countries are doing, is simply leading to food shortages, starvation, and even death in the long run. Among countries chasing a high ESG score, it is common to erect solar panels on prime farmland, which makes the food supply and farming worse. This will cause food prices to soar, as the fields produce less food. This victimizes the most vulnerable among us.

Meanwhile, food scarcity propaganda still persists. For instance, a simple search online will show the narrative that temperature increases or climate change will hurt crop yields. What we've learned in this chapter, however, is that plants grow better with more CO_2, fertilizers and longer growing seasons. They tolerate drought better with more CO_2, and they thrive because they use water more efficiently, which also helps their roots grow deeper. They grow bigger faster, and provide us with higher nutrient density from the crops we consume.

Meanwhile, academics are funded to make baseless claims which are the exact opposite of observational evidence.

> https://www.pnas.org › content
> **Temperature increase reduces global yields of major crops in ...**
> by C Zhao · 2017 · Cited by 932 — Without CO2 fertilization, effective adaptation, and genetic improvement, each degree-Celsius increase in **global** mean **temperature** would, on ...
>
> https://theconversation.com › climate-change-is-affectin...
> **Climate change is affecting crop yields and reducing global food**
> Jul 9, 2019 — As **climate change** alters temperature and rainfall patterns, **yields** of some **crops** are increasing while others decline. The net result: less ...
>
> https://climate.nasa.gov › news › global-climate-change...
> **Global Climate Change Impact on Crops Expected Within 10 ...**
> Nov 2, 2021 — **Climate change** may affect the **production** of maize (corn) and wheat as early as 2030, according to a new NASA study.
>
> https://www.nationalgeographic.com › crops
> **5 Ways Climate Change Will Affect You: Crop Changes**
> **Climate** alone doesn't dictate **yields**, political shifts, **global** demand, and **agricultural** practices will influence how farms fare in the future.
>
> https://www.sciencedirect.com › science › article › pii
> **Global vulnerability of crop yields to climate change - Science ...**
> by IS Wing · 2021 · Cited by 1 — In the absence of additional margins of adaptation beyond those pursued historically, projections constructed using an ensemble of 21 **climate** model ...
>
> https://www.greenbiz.com › article › climate-change-aff...
> **Climate change is affecting crop yields and reducing global ...**
> Jul 12, 2019 — Our study showed that **climate change** is **reducing** consumable food calories by around 1 percent yearly for the top 10 global **crops**. This may sound ...

Notice the very top is a study from a "scientific" website, PNAS, Proceedings of the National Academy of Sciences.

Among the many traumatizing psychological tactics used by climate alarmists and the climate change cult, primarily egged on by organizations such as the UN and WEF, is "gaslighting." Gaslighting is a practice by which an abuser tells their victim that they are crazy for stating an obvious fact or a legitimate feeling the victim has. In relationships, someone who is gaslighting will tell you that you're faking pain or overdramatizing when something hurts you. In the cult of climate change, the gaslighter is focused on making you look away from the truth about food production and yields, while the mainstream media promotes lies about carbon, weather, and temperatures. The gaslighter will tell you that their false reality will be safer, healthier, and happier.

By demonizing CO_2, the globalists are not just trying to downsize our fossil fuel usage; eventually they'll try to downsize *you and our families*.

Part Three:
CLIMATE LIES

CHAPTER EIGHT:
THE CARBON CYCLE UNVEILED

> *"The carbon cycle is Nature's way of recycling carbon atoms.*
> *It is the foundation for all life on Earth[44]."*
> National Oceanic and Atmospheric Agency (NOAA)

I'm a climate "heretic" because I believe that attacking CO_2 is attacking the foundation of all life on Earth and the climate cult thinks just the opposite. The Earth Observatory with NASA states:

> *Carbon is the backbone of life on Earth. We are made of carbon, we eat carbon, and our civilizations—our economies, our homes, our means of transport—are built on carbon. We need carbon, but that need is also entwined with one of the most serious problems facing us today: global climate change*[45].

At least NASA's quote is half accurate. Carbon is the backbone of life on Earth. Our civilizations and economies depend on it. Despite the catastrophic claims, we have no proof that there will ever be a climate disaster based upon rising CO_2 levels, which was the basis for Part Two of this book.

In Part Three, we'll look at how carbon dioxide and other factors affect the weather and what, if anything, we need to actually worry about.

The Earth naturally produces twenty to fifty times more carbon dioxide than what humans create through activities like burning fossil fuels, cement and steel production, and agriculture. That's right, the Earth is logically the

---
[44] https://oceanservice.noaa.gov/facts/carbon-cycle.html Accessed 5/20/23.
[45] https://www.earthobservatory.nasa.gov/features/CarbonCycle

number one culprit on the WEF and UN hit lists. Termites and decaying trees release far more CO_2 than man does.

The predominant source of CO_2 in our atmosphere is naturally occurring, not man made. The carbon dioxide that is emitted, whether through natural or man-made origin, is constantly cycled in a process of release and reabsorption by the Earth. In fact, estimates suggest that the Earth reabsorbs 100% of its self-generated CO_2 and about 50% of the CO_2 released by human activities.

The University of Colorado's Ashley Ballantyne and NOAA's Pieter Tans and their colleagues estimate that overall, oceans and natural ecosystems continue to pull about half of people's carbon dioxide emissions out of the atmosphere as well as all-natural CO_2 being released. Because the Earth is greener, there is an ever-growing natural production of more CO_2. Since emissions of CO_2 have increased substantially since 1960, Ballantyne said, "Earth is taking up twice as much CO_2 today as it was 50 years ago[46]". Remember, the great circle of life? There is more green life and plankton now than before we started releasing CO_2 from coal, oil, natural gas, and fossil fuels[47].

The Earth absorbs more human-generated CO_2 than in previous decades; some believe it is because we have a greener Earth and healthier plants (as evidenced by higher crop yields in Part Two).

One reason the Earth absorbs more CO_2 than in the past is plankton. Plankton, similar to land plants, plays a crucial role in recycling most of the atmospheric CO_2. Plankton absorbs and sequesters CO_2. And when plankton dies and sinks to the ocean floor, it takes its CO_2 with it.

Land plants and trees also contribute to the carbon cycle by absorbing CO_2 during their growth, and releasing it when they decompose. Interestingly, termites and bacteria play a significant role in this cycle, producing more CO_2 than human activities.

---

[46] https://phys.org/news/2012-08-earth-absorbing-carbon-dioxide-emissions.html
[47] https://www.earthobservatory.nasa.gov/features/CarbonCycle

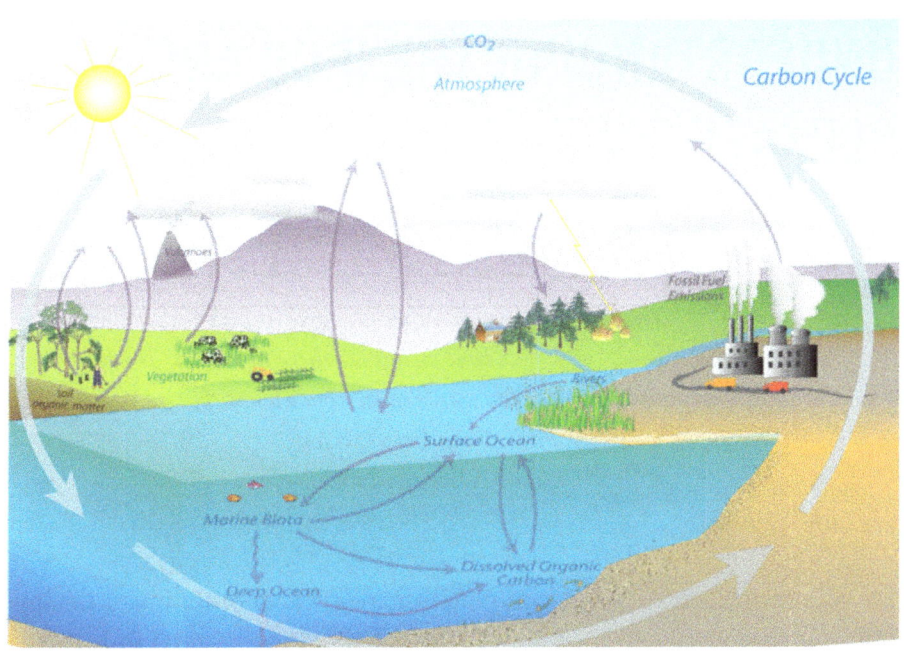

https://www.noaa.gov/education/resource-collections/climate/carbon-cycle
The carbon cycle - Image credit: NOAA

Part of the drum that the cult of climate change beats is the idea that increased carbon dioxide in the atmosphere is deadly. In Part Two of this book, I will illustrate that all life benefits from an increase in CO_2. In fact, indoor CO_2 levels can be as high as 1,000 or even 1,500 ppm, particularly in crowded spaces. Despite this, CO_2 remains a harmless gas that is essential for plant growth. It contributes only minimally to the warming effect when compared to water vapor and clouds.

Contrary to claims of harm, breathing 10,000 ppm of CO_2 is considered safe, as evidenced by sailors on submarines who regularly breathe even higher concentrations for extended periods. Unfortunately, today's narrative leads us to believe CO_2 is harmful.

Now that you have a grasp on basic CO_2 and carbon science, let's look deeper at the climate "science" and how the Intergovernmental Panel on Climate Change (IPCC) distorts the truth in order to scare people into compliance with a false reality.

UNDERSTANDING THE IPCC AND CLIMATE SCIENCE

> *"They knew that they were writing the constitution for hard times. They weren't writing it for easy times. Freedom of expression, I guarantee, is not for easy speech. It's not for the speech that we all agree on. It's the speech that nobody wants to hear. You know, that's what it's there to protect. The things that are unpopular. That nobody wants said. That's what that guarantee [the First Amendment] is for."*
> Robert F. Kennedy Jr.

Dr. Kiminori Itoh told the Intergovernmental Panel on Climate Change (IPCC) that warming fears are the "worst scientific scandal in history... When people come to know what the truth is, they will feel deceived by science and scientists."

Dr. Itoh earned his Ph.D in Industrial Chemistry from the University of Tokyo in 1978. He authored the Japanese book *Lies and Traps in the Global Warming Affair*. Dr. Itoh's significant academic contribution lies in the development of optical waveguide spectroscopy for solid surfaces for which he has received awards from relevant academic societies.

His interest in the global warming issue sparked around 1995 when he was asked to deliver a lecture on environmental meteorology. He questioned why the influence of solar changes had been largely overlooked by most climatologists. Since then, Dr. Itoh has authored or contributed to four books on this issue. He holds a patent on sunspot number anticipation and served as an expert reviewer for the IPCC AR4.

You might remember the IPCC from Part One of this book. The IPCC is a UN organization that plays a significant role in our public understanding of climate change. They shape narratives, though not always through honest science. What they commit is akin to a crime against humanity. While the IPCC provides some valuable assessments, their singular focus on human-induced climate change deliberately shrouds the public from a full understanding of climate dynamics. They deliberately ignore the work of Dr. Itoh and other scientists.

On Wikipedia we can read that the IPCC has adopted its rules of procedure in the "Principles Governing IPCC Work." These state that the IPCC will assess:

- the risk of climate change caused by human activities;
- its potential impacts, and
- possible options for prevention.

The IPCC also has a statement called "The IPCC's Acknowledgment of Uncertainty," which states: "In climate research and modeling, we should recognize that we are dealing with a coupled non-linear chaotic system, and **therefore that the long-term prediction of future climate states is not possible.**" That phraseology is important, so let's break it down.

1. *"Coupled"*

First of all, when we say that a system is *"coupled,"* we mean that its various components are interconnected and influence each other. This interdependence creates feedback loops, both positive and negative, which can dampen or amplify the system's behavior.

2. *"Non-Linear"*

A *"nonlinear system"* exhibits outputs that are not directly proportional to changes in inputs. In similar terms, small changes in inputs can lead to disproportionately large effects on outputs. For example, slight alterations in cloud cover can have a substantial impact on weather patterns, regardless of CO_2 levels. Similarly, a significant change in one part of the system may result in minor effects elsewhere, illustrating the complexity of interconnected systems.

Natural systems often exhibit nonlinear behavior, characterized by variables that evolve over time in unpredictable and sometimes chaotic ways. This unpredictability underscores the challenges of accurately predicting outcomes in complex systems.

The UN's IPCC on Climate Models,
(IPCC third Assessment Report (2001) Section 14.2.2.2, page 774):

"In climate research and modeling, we should recognize that we are dealing with a coupled non-linear chaotic system, and therefore that the long-term prediction of future climate states is not possible."

Understanding these principles is crucial for comprehending the dynamics of natural phenomena and how they may respond to various influences over time. Natural systems have more negative feedback loops than positive ones. Natural systems seek equilibrium. Any change has countervailing forces to balance them.

This means that so-called "tipping points" or runaway global warming is highly unlikely. If I were a dishonest propagandist, I'd say that global warming could never happen with complete certainty. But I wouldn't make such a sweeping, binary statement because it's irresponsible. However, climate alarmists make such statements continually, even in the presence of scientific data that strongly suggests otherwise, even just the opposite.

No one can predict the future with certainty. In the eons of Earth, even in the last 6,000 years of recorded history, we have seen very abrupt changes in climate and weather, completely unrelated to CO_2 or water vapor. The Bible tells us of seven-year famines, great floods, and other destructive weather events.

Unfortunately, climate science and the influence of organizations like the IPCC are highly politicized. Their reports are manipulated for political agendas rather than presenting unbiased scientific findings. IPCC summary documents are usually, if not always, released months before the full report, so they cannot be challenged. Worst of all, these reports are written at the direction of bureaucrats and government officials, *not scientists*. There is no way to verify nor deny these snippets that are released early until the full thousand-page reports are made public.

However, very few people read the full reports. Mainstream media does not cover inaccuracies nor follow up with fact checking. Due to this rampant ignorance and failure in journalism (which is on purpose), IPCC reports contain inaccuracies that are never corrected before they're passed on, or even sensationalized, for public consumption.

In the same way that the Covid-19 event saw a proliferation of brave (and highly censored) scientists and healthcare professionals who are just now starting to make waves in the courts and put pressure on their local government officials to "apologize" for spreading misinformation and mandates unlawfully, there are some courageous voices in the scientific community who have called for a reevaluation of the IPCC's role and funding. The dishonest approach detailed above clearly hinders scientific progress and leads to ineffective, unnecessary, and expensive policy measures.

Because the IPCC promotes the assumption (which has little to no basis in real science) that climate change is nearly all man-made, and some real scientists are calling them out as the shills and false prophets. They are only fueling the negative aspects of the Earth's warming or the reason for this warming being man made. **They are paid to do this.** Those scientists who disagree are not invited back and are marginalized.

"Trust the science" is a phrase that some say was intentionally planted as a method of mind control by the CIA. It's an interesting theory when you apply it to climate change. Between the media, former President Barack Obama, and the IPCC, our environment is a veritable storm of supposed authority figures telling us to "trust them" while they feed us proven lies.

But don't just listen to me, here are a handful of quotes from actual scientists[48]...these quotes all came from a 2007 Senate Testimony. This list isn't comprehensive, as there are many other notable and credible scientists who have expressed dissenting views. If I had it my way, I'd hang these quotes up in every classroom across America--and the world--so that children would have a sliver of hope in hearing the truth about climate change; so, they know that true science is done with debate at its core, dissecting and testing

---

[48] https://www.epw.senate.gov/public/index.cfm/press-releases-all?ID=2158072e-802a-23ad-45f0-274616db87e6

different theories until one emerges as the prevailing theory; similar to Einstein's Theory of Relativity, which was mentioned earlier in this book.

TESTIMONIES FROM THE 2007 U.S. SENATE ENVIRONMENT AND PUBLIC WORKS COMMITTEE MINORITY STAFF REPORT

> 1. *"The IPCC has actually become a closed circuit; it doesn't listen to others. It doesn't have open minds... I am really amazed that the Nobel Peace Prize has been given on scientifically incorrect conclusions by people who are not geologists."*

Indian geologist Dr. Arun D. Ahluwalia at Punjab University. A board member of the UN-supported International Year of the Planet.

> 2. *"The models and forecasts of the UN IPCC are incorrect because they are only based on mathematical models and presented results at scenarios that do not include, for example, solar activity."*

Victor Manuel Velasco Herrera, a researcher at the Institute of Geophysics of the National Autonomous University of Mexico

> 3. *"It is a blatant lie put forth in the media that makes it seem there is only a fringe of scientists who don't buy into anthropogenic global warming."*

U.S Government Atmospheric Scientist Stanley B. Goldenberg of the Hurricane Research Division of NOAA (The National Oceanic and Atmospheric Administration)

> 4. "Even doubling or tripling the amount of carbon dioxide will virtually have little impact, as water vapor and water condensed on particles as clouds dominate the worldwide scene and always will."

Geoffrey G. Duffy, a professor in the Department of Chemical and Materials Engineering of the University of Auckland, NZ

> 5. "Since I am no longer affiliated with any organization nor receiving any funding, I can speak quite frankly...As a scientist I remain skeptical."

Atmospheric Scientist Dr. Joanne Simpson, the first woman in the world to receive a PhD in meteorology, formerly of NASA who has authored more than 190 studies and has been called "among the most preeminent scientists of the last 100 years."

> 6. "Creating an ideology pegged to carbon dioxide is a dangerous nonsense... The present alarm on climate change is an instrument of social control, a pretext for major businesses and political battle. It became an ideology, which is concerning."

Environmental Scientist Professor Delgado Domingos of Portugal, the founder of the Numerical Weather Forecast group, has more than 150 published articles.

> 7. "CO2 emissions make absolutely no difference one way or another.... Every scientist knows this, but it doesn't pay to say so...Global warming, as a political vehicle, keeps Europeans in the driver's seat and developing nations walking barefoot."

Dr. Takeda Kunihiko, Vice-Chancellor of the Institute of Science and Technology Research at Chubu University in Japan

8. *"The [global warming] scaremongering has its justification in the fact that it is something that generates funds."*

Award-winning Paleontologist Dr. Eduardo Tonni, of the Committee for Scientific Research in Buenos Aires and head of the Paleontology Department at the University of La Plata, Argentina

9. *"The cause of these global changes is fundamentally due to the Sun and its effect on the Earth as it moves about in its orbit. Not from man-made activities."*

Retired Award-Winning NASA Atmospheric Scientist Dr. William W. Vaughan, recipient of the NASA Exceptional Service Medal, a former Division Chief of NASA's Marshall Space Flight Center and author of more than 100 journal articles, monographs, and papers

10. *"Unfortunately, Climate Science has become Political Science…It is tragic that some perhaps well-meaning but politically motivated scientists who should know better have whipped up a global frenzy about a phenomena which is statistically questionable at best."*

Award-Winning Princeton University Physicist Dr. Robert H. Austin, who has published 170 scientific papers, was elected a member of the U.S. National Academy of Sciences and is the current Chair of the U.S. Liaison Committee of the International Union of Pure and Applied Physics. Austin also won the 2005 Edgar Lilienfeld Prize of the American Physical Society.

In 2010, more dissenting voices were added to the U.S. Senate Committee on Environmental and Public Works testimony database, which include the following:

> 1. *"Please remain calm: The Earth will heal itself — Climate is beyond our power to control...Earth doesn't care about governments or their legislation. You can't find much actual global warming in present-day weather observations. Climate change is a matter of geologic time, something that the earth routinely does on its own without asking anyone's permission or explaining itself."*

Nobel Prize-Winning Stanford University Physicist Dr. Robert B. Laughlin, who won the Nobel Prize for physics in 1998, and was formerly a research scientist at Lawrence Livermore National Laboratory.

> 2. *"The energy mankind generates is so small compared to that overall energy budget that it simply cannot affect the climate...The planet's climate is doing its own thing, but we cannot pinpoint significant trends in changes to it because it dates back millions of years while the study of it began only recently. We are children of the Sun; we simply lack data to draw the proper conclusions."*

Russian Scientist Dr. Anatoly Levitin, the head of Geomagnetic Variations Laboratory at the Institute of Terrestrial Magnetism, Ionosphere and Radiowave Propagation of the Russian Academy of Sciences.

> 3. *"Hundreds of billion dollars have been wasted with the attempt of imposing a Anthropogenic Global Warming (AGW) theory that is not supported by physical world evidences...AGW has been forcefully imposed by means of a barrage of scare stories and indoctrination that begins in the elementary school textbooks."*

Brazilian Geologist Geraldo Luís Lino, who authored the 2009 book *The Global Warming Fraud: How a Natural Phenomenon Was Converted into a False World Emergency*.

> 4. "The dysfunctional nature of the climate sciences is nothing short of a scandal. Science is too important for our society to be misused in the way it has been done within the Climate Science Community." The global warming establishment "has actively suppressed research results presented by researchers that do not comply with the dogma of the IPCC."

Swedish Climatologist Dr. Hans Jelbring, of the Paleo Geophysics & Geodynamics Unit at Stockholm University. [Updated December 9, 2010. Corrects Jelbring's quote.]

> 5. "Global warming is the central tenet of this new belief system in much the same way that the Resurrection is the central tenet of Christianity. Al Gore has taken a role corresponding to that of St. Paul in proselytizing the new faith...My skepticism about AGW arises from the fact that as a physicist who has worked in closely related areas, I know how poor the underlying science is. In effect the scientific method has been abandoned in this field."

Atmospheric Physicist Dr. John Reid, who worked with Australia's CSIRO's (Commonwealth Scientific and Industrial Research Organization) Division of Oceanography and worked in surface gravity waves (ocean waves) research.

> 6. "We maintain there is no reason whatsoever to worry about man-made climate change, because there is no evidence whatsoever that such a thing is happening."

Greek Earth scientists Antonis Christofides and Nikos Mamassis of the National Technical University of Athens Department of Water Resources and Environmental Engineering

> 7. "The whole idea of anthropogenic global warming is completely unfounded. There appears to have been money gained by Michael Mann, Al Gore and UN IPCC's Rajendra Pachauri as a consequence of this deception, so it's fraud."

South African astrophysicist Hilton Ratcliffe, a member of the Astronomical Society of Southern Africa (ASSA) and the Astronomical Society of the Pacific and a Fellow of the British Institute of Physics.

> 8. *"Let's be clear: the work of science has nothing whatever to do with consensus [which] is the business of politics. . . . What is relevant is reproducible results. The greatest scientists in history are great precisely because they broke with the consensus."*

Atmospheric Scientist Timothy R. Minnich, who has more than 30 year's experience in the design and management of a wide range of air quality investigations for industry and government, is a past member of the American Meteorological Society and specializes in issues like acid rain and ozone and has authored or co-authored numerous technical publications and reports.

> 9. *"Based on the laws of physics, the effect on temperature of man's contribution to atmospheric CO2 levels is minuscule and indiscernible from the natural variability caused in large part by changes in solar energy output."*

Atmospheric Scientist Robert L. Scotto, who has more than 30 years air quality consulting experience, served as a manager for an EPA Superfund contract and is co-founder of Minnich and Scotto, Inc., a full-service air quality consulting firm. He also is a past member of the American Meteorological Society (AMS). Scotto, a meteorologist who has authored and co-authored numerous technical publications and reports.

> 10. *"Any reasonable scientific analysis must conclude the basic theory wrong!!"*

NASA Scientist Dr. Leonard Weinstein who worked 35 years at the NASA Langley Research Center eventually becoming a Senior Research Scientist. He is presently a Senior Research Fellow at the National Institute of Aerospace.

CHAPTER NINE:
CLIMATE MODELS AREN'T READY FOR PRIME TIME

"[The models have] no understanding of cloud formation/forcing."
"Assumptions are made, then adjustments are made to support a narrative. Our models are mickey-mouse mockeries of the real world."
Dr. Mototaka Nakamura

Dr. Mototaka Nakamura received a Doctorate of Science from the Massachusetts Institute of Technology (MIT), and for nearly 25 years specialized in abnormal weather and climate change at prestigious institutions that included MIT, Georgia Institute of Technology, NASA, Jet Propulsion Laboratory, California Institute of Technology, JAMSTEC, and Duke University.

In his book *The Global Warming Hypothesis is an Unproven Hypothesis,* Dr. Nakamura explains why the data foundation underpinning global warming science is "untrustworthy" and cannot be relied on:

https://electroverse.info/climate-scientist-breaks-ranks/

Modeling the climate is a difficult task. There are a multitude of variables that affect our climate. One might say that they have a lot of problems. They're not testable because they haven't been around long enough to compare real temperatures with the model results. What's more, there are dozens of climate models.

Nature is continually changing; it has cycles in periods of months, years, centuries, and even millennia. Some variables in climate models include the sun, which often provides just a little bit more or less energy. Planets are a variable, as well as ocean currents and the El Nino or La Nina effect, also known as ENSO.

ENSO represents the irregular unpredictable changes of the Pacific Ocean from warmer to cooler. This has a large effect on the climate across the world, causing more snow and rain in some areas and higher or lower temperatures in other areas. ENSO is always changing. Although nobody can explain why, it does seem that after each warm El Nino, we have a slight step up in temperature.

Dr. Roy W. Spencer is a prominent figure in climate science, serving as a Principal Research Scientist at the University of Alabama in Huntsville (UAH). With a background as a meteorologist, Spencer previously held the position of Senior Scientist for Climate Studies at NASA. His research spans a wide range of weather and climate topics.

One of Spencer's notable contributions is his co-development, alongside John Christy, of the original method for monitoring global atmospheric temperature variations using Earth-orbiting satellites. This groundbreaking work earned him recognition, including NASA's Medal for Exceptional Scientific Achievement and the American Meteorological Society's Special Award. During his tenure at NASA, Spencer also served as the U.S. Science Team Leader for the Advanced Microwave Scanning Radiometer (AMSR-E) aboard NASA's Aqua satellite.

At the University of Alabama in Huntsville, Spencer's research focuses on understanding the sensitivity of the climate system to increasing greenhouse gas concentrations and investigating the impact of the urban heat island on land surface air temperature trends. Additionally, he maintains and updates the satellite record of global atmospheric temperatures monthly on his website, www.drroyspencer.com. He's also an accomplished author, having published several books on climate change, including the New York Times bestseller, *Climate Confusion*.

Dr. Spencer has contributed a graph that reinforces the fact that climate models are highly inaccurate, overestimating warming. One of the many problems is that you cannot model clouds. Clouds are naturally important to the Earth's temperature. They cool the Earth during the day by reflecting the sun's energy and warm the Earth during the night by holding in the days' warmth. Remember, Dr. Will Happer told us water vapor and clouds make up 95% of the greenhouse effect.

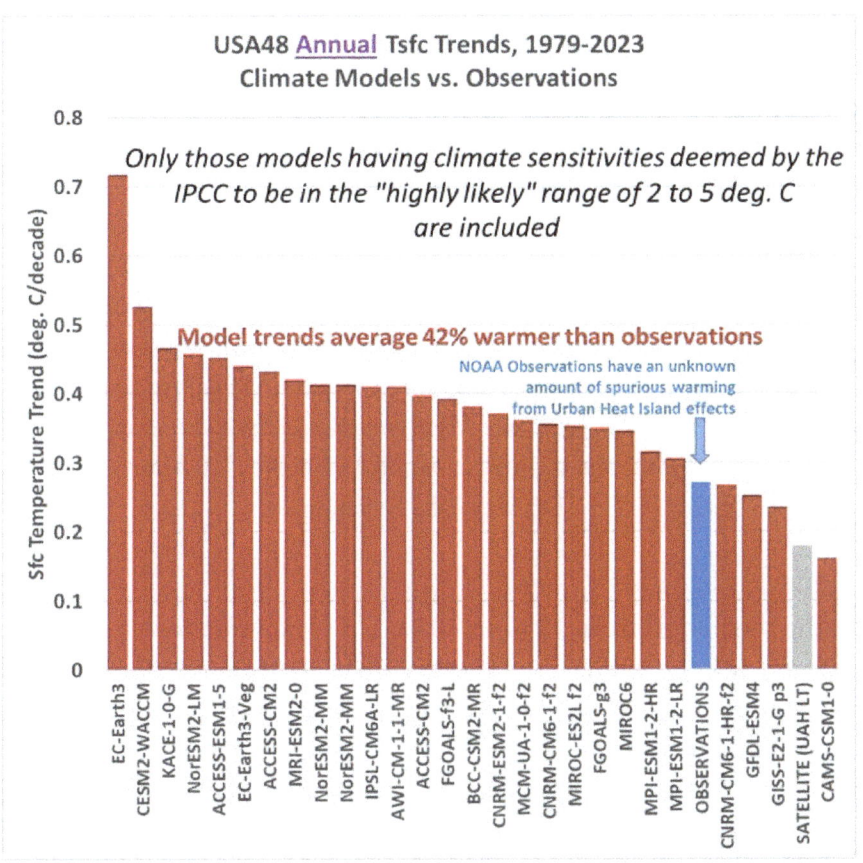

U.S.A. Temperature Trends, 1979-2023: Models vs. Observations
February 2nd, 2024, by Roy W. Spencer, PHD

In a graph titled "USA Temperature Trends, 1979–2023," Spencer estimates a decade warming trend on the left-hand side. The models are designed to predict warming caused from increasing CO_2. The graph doesn't make it clear whether a comparison is being made to the adjusted or raw temperatures, which means the actual temperature shown by the blue bar should be less. The graph implies that in a hundred years we'll see ten times as much warming as is being observed. So, their model predictions get worse with time because they are compounded.

In case it isn't blatantly obvious, climate modeling needs a lot more work. Yet, policy makers and propagandists are quick to tell us they are accurate and use them for policy and propaganda.

GLOBAL TEMPERATURES

> *"The data foundation underpinning global warming science is untrustworthy and cannot be relied on…Global mean temperatures before 1980 are based on untrustworthy data. Before full planet surface observation by satellite began in 1980, only a small part of the Earth had been observed for temperatures with only a certain amount of accuracy and frequency. Across the globe, only North America and Western Europe have trustworthy temperature data dating back to the 19th century[49]."*
>
> Dr. Mototaka Nakamura

During the period from 1990 to 2014, Dr. Nakamura focused on cloud dynamics and the interaction of forces affecting atmospheric and oceanic flows at scales ranging from medium to planetary. He has contributed over 20 climate papers on fluid dynamics, showcasing his expertise and depth of knowledge in the field. Dr. Nakamura's credibility and extensive knowledge base are widely acknowledged within the scientific community.

Weather and climate are closely related concepts, with weather representing short-term atmospheric conditions and climate referring to long-term patterns. Weather and climate are in constant flux, always changing regardless of human intervention or control. Temperature variations are common, too. Temperature fluctuations of 5, 10, and 15 degrees in a single day are normal. Sudden shifts, such as going from warm to freezing temperatures, are not unusual either. A friend of mine in Colorado often says, "If you don't like the weather in Colorado, wait ten minutes."

---

[49] https://electroverse.info/climate-scientist-breaks-ranks/

In the past, major shifts in temperature have been recorded. For instance, in 1877–1878, Minnesota didn't have a winter at all! That's right, it was a warm winter in Minnesota—the opposite of "hell freezing over." Conversely, there were a couple of years without summer because of a volcanic eruption in Indonesia in 1816[50,51]. Hot and cold weather events have been common throughout recorded history. The weather and temperature of Earth has always changed and always will.

Geographical location also plays a significant role in weather patterns. For instance, Hawaii experiences relatively consistent warm weather but exhibits diverse climates across the islands due to mountainous terrain, which influences rainfall distribution. If you go to Hawaii and it is raining, you can almost always drive somewhere on the same island that has sun.

You've probably experienced dramatic shifts in weather firsthand, such as driving into or out of storms. Regions near the North and South poles generally remain cold, but occasional warming events occur, often propagandized by the media which focuses on the alarming aspect of ice caps melting, wildlife overheating, or the ocean rising.

These natural, common, and unalarming phenomena are sensationalized by soundbites such as, "If Greenland continues to melt, Florida will be underwater." These headlines and soundbites don't take into account that the ice freezes again that night and may stay frozen for weeks after that single warm day. What's more, there simply isn't enough ice in Greenland to increase sea levels enough to affect people in any surrounding area, much less in Florida.

In the current society of climate change propaganda, cold events are essentially ignored. Cleverly, some media outlets are telling us that cold weather events are caused by warming, too, in order to divert our attention from anything having to do with common sense. The climate change narrative doesn't need logic nor does it check facts.

Seasonal variations further influence weather patterns, evident in phenomena like rainy seasons, monsoons in tropical areas and distinct seasons in deserts. Consider the vast difference between winter and summer in Canada.

---

[50] https://newenglandhistoricalsociety.com/1816-year-without-a-summer/
[51] https://www.almanac.com/year-without-summer-mount-tambora-volcanic-eruption

Determining a global average temperature isn't possible due to the Earth's vast size and climatic diversity. Historical temperature records don't exist with any accuracy, yet they are presented as facts.

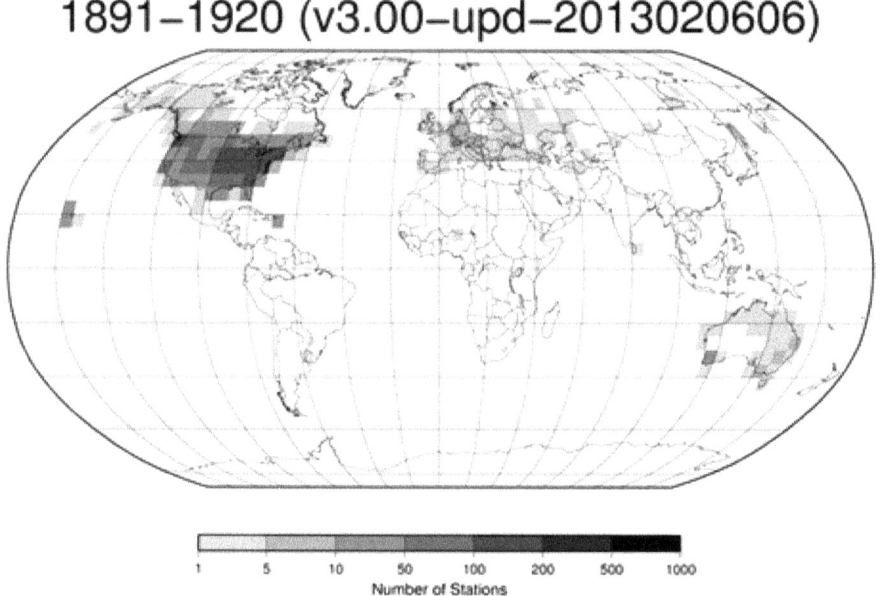

Stations 1891 to 1920 world pic

In the graph above you can see where the Earth had temperature measurements between 1891 and 1920. Outside of the US, Europe, and Australia there were almost no temperature records all. There is no way to know the worldwide temperatures during this time. The data doesn't exist. In order to take these temperatures, a person had to physically visit a thermometer and record readings in a journal. Naturally, unless the temperatures were taken at the same time of day for a year straight these numbers lack scientific accuracy. In fact, I hesitated to put this graph in the book because it requires identifying colors to read, but since the graph's information is so inherently flawed, it doesn't matter whether you read it...I am mostly pointing out the flawed science that is often used as fact to substantiate whatever narrative is desired.

We don't have much data about temperatures before about 1960. In fact, collecting weather data was difficult during World War II (1940–1945), and we have almost no record from the North and South poles, Asia, Africa, South America, and the oceans.

This presents one of the myriads of problems we have to create valid climate models: we don't have reliable historical temperature records. Period.

We didn't know what the Earth's temperature was in the years before 1960, and we still don't.

Are you starting to see a pattern here?

The climate establishment bases a world of political decisions, regulations, and propaganda campaigns based off of historical temperature records that *don't exist*.

What's more, the average temperature is relatively meaningless. Let's take the moon as an example. The temperature on the moon can reach a blistering 250° Fahrenheit (120° Celsius) during lunar daytime at the moon's equator, and plummet to -208 degrees F (-130° C) at night. The temperatures are even colder at the moons' poles. The average temperature of the moon without the poles would be 23°F (-5° C). At the poles it would be even colder.

How much colder? How could that be determined accurately? How would we establish a reasonable and accurate average temperature of the moon? What would happen to that average if we started with measurements in some areas of the moon and then added thousands more temperature stations all over the moon? What if we took readings of both the hot as well as cold areas? How would we be able to know they were accurate over time? It doesn't matter because it isn't even possible. Even if we had a journal of long-term readings from pristine temperature stations, unless we continue getting actual readings from the moon, we'll begin to rely on simulated data. The record would be compromised and accuracy, impossible.

However, this is exactly what NOAA is doing with temperature stations on Earth. Like the Earth, the Moons' temperature numbers are meaningless. The average temperature doesn't tell you what clothes to bring for the cold or the heat. What would be the real average, to a tenth of a degree? If it were possible, it wouldn't have any real meaning. Taking an average temperature on a place as diverse as the Earth would require endless samples in tens of thousands of locations over a long period of time, at least decades, if not centuries, before a climate model could even be remotely considered credible. However, because weather and temperature cycles on Earth don't appear to run in decades, but in centuries, this would still be a grossly flawed approach to climate modeling.

If I were a scientist getting paid to present a certain urgent "bend" on my work, such as the philosophy that the Earth is heating up, I might be consciously or unconsciously tempted to place my temperature reading stations in hotter places. If I were the corporate entity or university funding the research, I might be tempted to do the same. If you think this doesn't happen, simply read the testimonies from the 2007 U.S. Senate Hearing on the environment where over 600 scientists submitted evidence that they don't merely doubt climate change, but many of them thought they were manipulated into supporting it while on staff at their respective university or institution.

Climate scientists use a mere 30-year temperature average as a benchmark for assessing present day changes in the absence of historical data. However, as we saw in the beginning of this book, just a few decades ago they feared another Ice Age. Additionally, records from remote regions like the poles or oceans, are sparse and unreliable. Even today, vast areas of Africa, South America, and Asia do not have any measurements at all in the present, let alone historical data.

While efforts are made to understand and predict climate patterns, the complexity of Earth's climate system and historical data limitations make it impossible to have a temperature record as accurate or meaningful as scientists and climate alarmists claim to have.

In conclusion, global average temperature is meaningless and is not built on real world data.

CHAPTER TEN:
CLIMATE FAKE NEWS

As you've seen, climate alarmists use a variety of tactics to manipulate data, making a climate change mountain out of a molehill. In the graph below you'll see two pictures, one that looks like a stock chart trending upward, the other looks like a flat red band with a barely discernible gradual increase. Interestingly, they represent the same numbers. The chart on the left appears to show a meteoric rise because it only shows a tiny sliver of the data. The one on the right shows that data in a proportionate way over time. Same data, looked at very differently.

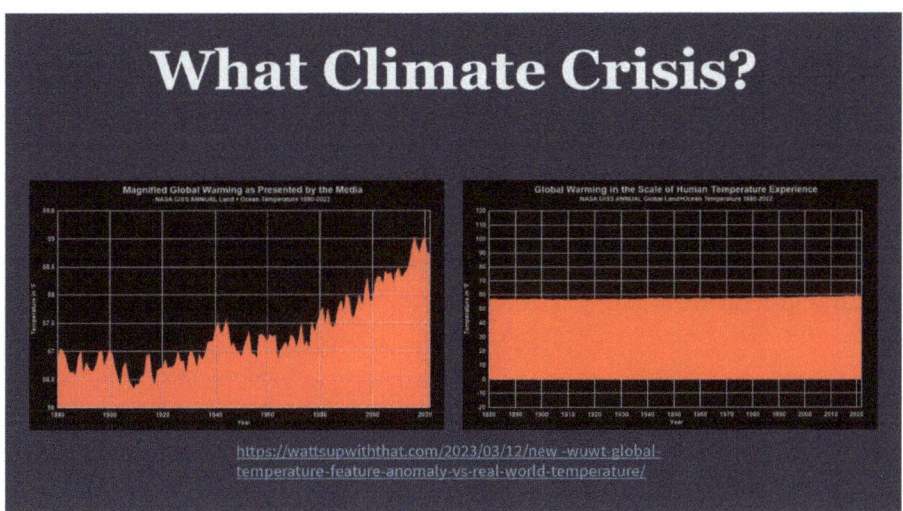

The next chart illustrates "Contiguous U.S. Average Temperature" based upon monthly averages since 1895. We just covered how this temperature averaging is an inherently flawed process anyway, but you don't see anything that looks alarming nor out of line, do you?

As you might have already guessed, the data is never presented to us this way. More frighteningly, I have pulled these numbers repeatedly and found that, mysteriously, the historical data continues to "evolve" on a regular basis. I'm not sure if our scientists are time traveling, but something smells fishy.

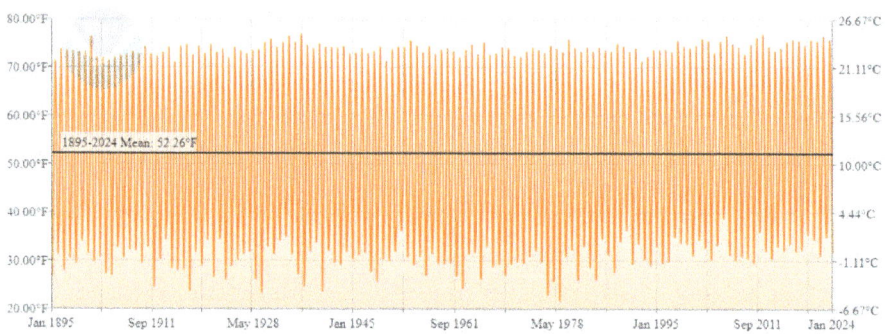

NOAA Climate at a Glance National Time Series[52]
No filter, one-month average temperature, all months for 1895 to 2024, accessed 2/15/2024
https://www.ncei.noaa.gov/access/monitoring/climate-at-a-glance/national/time-series/110/tavg/1/0/1895-2024?base_prd=true&begbaseyear=1895&endbaseyear=2024

The next graph was accessed at the same time, ending in December of 2023 versus January of 2024. In this NOAA chart, the mean temperature for January was 52.26 degrees Fahrenheit and it was 52.28 degrees Fahrenheit for the graph ending in December 2023. One month should not change the mean temperature at all.

---

[52] https://www.ncei.noaa.gov/access/monitoring/climate-at-a-glance/national/time-series/110/tavg/1/0/1895-2024?base_prd=true&begbaseyear=1895&endbaseyear=2024

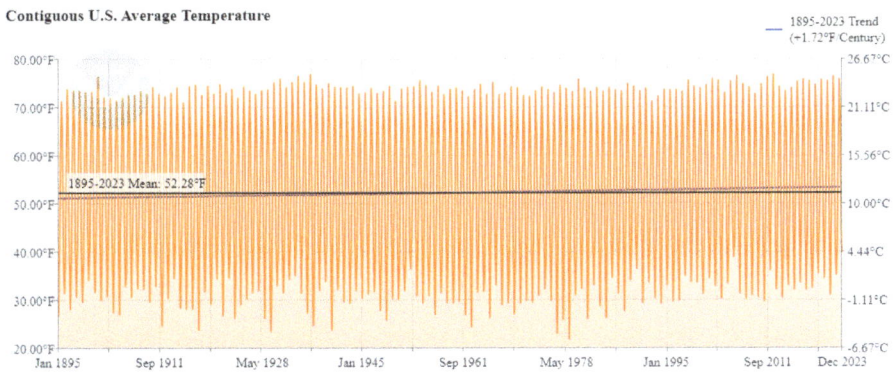

NOAA Climate at a Glance National Time Series[53]
No filter, one-month average temperature, all months for 1895 to 2023, accessed 2/15/2024

More importantly, notice how the high temperatures at the top are not getting higher. We are not getting warmer, as the claim goes. In fact, we are actually getting milder winters. I'm not sure about you, but living in Wisconsin, I'd welcome a milder winter, myself.

When you dig into the data as I have, you come to find out that our days are not as warm as they used to be and our nights aren't as cool as they used to be. This means that when you average the high of the day and the low of the day you can have the same temperature even in milder climates where the temperature doesn't have as dramatic swings, which is considered a "milder" climate. In a milder climate, the daytime temperature isn't as warm and the nighttime temperature isn't as cool.

Although the Secretary General of the UN says data coming out of all major media and universities would agree that "we're boiling," the truth is that we are experiencing a milder climate than in previous years.

---

[53] https://www.ncei.noaa.gov/access/monitoring/climate-at-a-glance/national/time-series/110/tavg/all/7/1895-2023?base_prd=true&begbaseyear=1895&endbaseyear=2023&trend=true&trend_base=100&begtrendyear=1895&endtrendyear=2023

CLIMATE CHANGE IS LOCAL, NOT PLANET-WIDE...

Dr. Dyson John Freeman, the Professor Emeritus at the Princeton Institute for Advanced Studies who was introduced earlier in this book, brings up another faulty misconception in the climate change narrative:

"THERE IS NO GLOBAL WARMING, THERE IS ONLY REGIONAL WARMING AND IT'S A GOOD THING!

The change that's now going on is very strongly concentrated in the Arctic. In fact, in three respects, it's not global, which I think is very important.

First of all, it is mainly in the Arctic.

Secondly, it's mainly in the winter rather than summer.

And thirdly, it's mainly in the night rather than at the daytime.

In all three respects, the warming is happening where it is cold, not where it is hot. The people in Greenland love it. They tell you it's made their lives a lot easier. They hope it continues. I am not saying none of these consequences are happening. I am just questioning whether they are harmful."

Over the past 175 years, the mild warming that we've experienced is uneven. The tropics have not warmed much at all, even though that's where most of the heat of the sun comes on to Earth and where the oceans evaporate a lot of water vapor, the most important greenhouse gas. That moist water-filled air then drifts towards the poles. Because the North Pole is warming more than the rest of the Earth, there is less difference in temperature between the warmer tropical air traveling north and the colder Arctic air. If the entire

Earth were warming, the tropics would be warming up commensurate with the Arctic, which they are not.

The South Pole has not been warming at all. In fact, it has experienced no warming in seventy years according to what data we have. Surprisingly, major news media have accurately reported that the South Pole has been setting cold records in the last few years.

Part of the climate change orthodoxy is that both poles should be warming. But the South Pole is not. This likely has to do with the tilt of the Earth and the South Pole not getting as much sunlight or sun's energy.

As Dr. Freeman pointed out, all weather and climate are local. I live a few miles from Lake Michigan, and we get Lake Michigan effect snow and temperatures. There can be several inches of snow difference between where I live and just 20 miles away. There can also be 10- or 20-degrees difference between the temperature where I live and just 20 miles away. Taking accurate temperature readings is darn near impossible for Wisconsin, let alone the entire Earth.

Pay close attention the next time you watch the weather on the news or a weather app. You will likely see a wide variety of temperatures within your state and just 20 or 50 miles away. My friend in Colorado keeps bookmarks on the town where her son goes to school: the area near the bus stop, which is fifteen miles away from the school, and their home which is 20 miles away from that. Depending on the grossly varied temperatures in that thirty-five-mile radius, playdates can range and activities determined by a fifteen-minute drive. Imagine trying to get an average temperature for your county, and then expand it and then expand it again to the whole world.

If I provided you with graphs of each individual state, and I've pulled many of them, you would see that even states that are right next to each other have different average temperatures. Consider the difference in temperature between Northern and Southern California or San Diego and Los Angeles.

Again, temperature and weather are always changing. They will continue to do so with or without more CO_2. Although there is plenty of evidence of warming, for the last 175 years we've also had a remarkably stable climate.

TEMPERATURE ISN'T AS ACCURATE AS THEY ARE TELLING YOU

We met Walter Cunningham earlier in this book, a man renowned as America's second civilian astronaut. During his eight-year tenure with NASA, he played a pivotal role in the design, development, and testing of all major operating systems for the Apollo spacecraft. In 1966, he served as a member of the prime crew for Apollo 2 and as part of the backup crew for Apollo 1.

In 1968, Walter achieved the remarkable feat of orbiting the Earth 163 times as the pilot of Apollo 7, marking the inaugural manned flight of the Apollo Program aimed at landing a man on the Moon. Notably, Apollo 7 stands as the longest, most ambitious, and most successful maiden flight of any manned vehicle to date. After the Apollo 7 mission, he assumed the role of Chief of the Skylab branch within the Astronaut Office.

Beyond his achievements in space exploration, he also boasts an illustrious military career as a Marine Corps fighter pilot, retiring with the rank of Colonel, USMCR. With over 4,500 hours of flying time, including 263 hours in space, he has demonstrated exceptional skill and dedication. He holds a master's degree in physics from UCLA and is an alumnus of the AMP Program at the Harvard Graduate School of Business.

In a 2013 interview titled, "A Conversation with Apollo Astronaut Walter Cunningham About a Vital Need To Restore Climate Science Integrity,[54]" Cunningham states:

> *"Those of us fortunate enough to have traveled in space bet our lives on the competence, dedication, and integrity of the science and technology professionals who made our missions possible...In the last twenty years, I have watched the high standards of science being violated by a few influential climate scientists, including some at NASA, while special interest opportunists have abused our public trust.*

---

[54] https://www.forbes.com/sites/larrybell/2013/08/06/a-conversation-with-apollo-astronaut-walter-cunningham-about-a-vital-need-to-restore-climate-science-integrity/?sh=3c99eb033345

> *The biggest problems I see with the sorry state of 'climate science,' as the public comes to know it through the media, are the alarmist claims, unsupported by data and history, being presented as facts.*
>
> *When these claims cannot be validated by empirical data, they attempt to justify them by equally dishonest claims of proof by 'consensus.'*
>
> *These alarmist claims create unwarranted fear in order to promote their political and profiteering agendas, while establishing regulatory policies that kill business and grow government--all at a terrific cost to taxpayers and energy consumers.*
>
> *We developed a letter to NASA Administrator Charlie Bolden and obtained signatures from seven Apollo astronauts, several former Headquarters managers, and Center directors, and 40 former management-level technical specialists.*
>
> *We asked that he restrain NASA from including unproven claims in public releases and on websites. Statements by NASA that man-made carbon dioxide was having a catastrophic impact on global climate change are not substantiated, especially when considering thousands of years of empirical data. It is clear that the science is NOT settled."*

Amusingly, *The Huffington Post* placed an Editor's Note at the bottom of the reprinted article that says, "We believe it's newsworthy when 49 former NASA scientists and astronauts pen a letter to the agency -- or to anyone -- about climate change. But what really raised temperatures is when we asked our readers to weigh in[55]..." If you're like me, you might have had the urge to suddenly slap your forehead and shake your head in surprise. Wait, let's hear what the NASA scientists have to say, and then let's troll them.

---

[55] https://www.huffingtonpost.co.uk/entry/nasa-global-warming-letter-astronauts_n_1418017

THE URBAN HEAT ISLAND EFFECT

The Urban Heat Island Effect (UHI) is a well-documented phenomenon during which urban areas experience higher temperatures compared to surrounding rural areas. As cities grow, they contribute to increased heat, affecting temperature measurements. This effect is significant and can distort temperature records, especially if adjustments are not made to account for it.

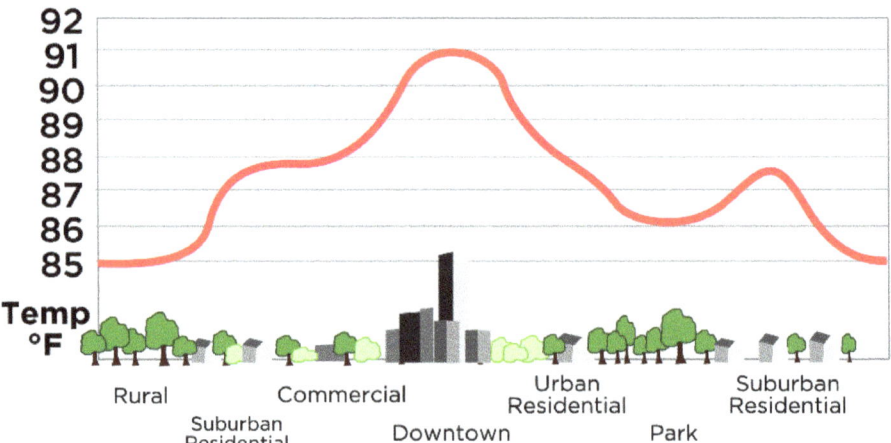

By TheNewPhobia[56]

Notice in this graph how a city has a six degree increase in temperature compared to the nearby rural area. The suburban area is warmer than the rural area too. This is the Urban Heat Island Effect that some studies have pointed out contributes about forty percent of the warming we see in measured temperatures. That is because the temperature station is located in an area that used to be rural and is now suburban or urban.

The location of weather stations is crucial for obtaining accurate temperature measurements, as we've established. Stations located in areas with minimal human influence, such as rural settings, provide more reliable

---

[56] https://commons.wikimedia.org/w/index.php?curid=5263244

data than those in urban or compromised environments. However, many weather stations have been affected by urbanization or changes in their surroundings over time, leading to warming biases in temperature records.

The following graphic illustrates how urban stations can be compromised. The temperature reading device is located next to many heat-causing buildings and structures, such as asphalt, exhaust fans, cell towers, and buildings. You can see how the temperature climbs as more heat-causing structures are located near it.

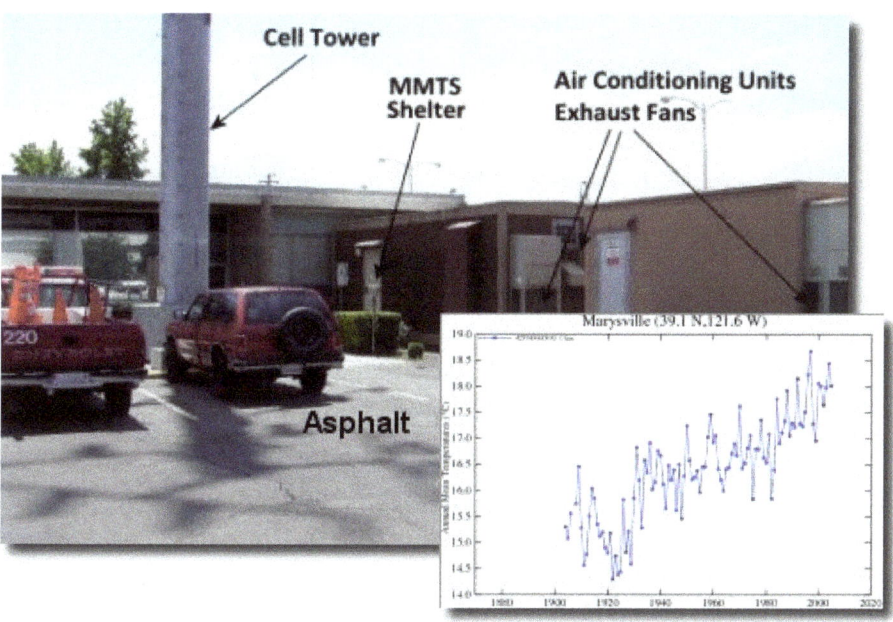

Graph courtesy of https://data.giss.nasa.gov/gistemp/ accessed 2/13/2023.

Compare this compromised reading to a well-maintained United States Historical Climatology Network (USHCN) station in Orlando, California. It is a properly situated weather station. It has not been compromised by heat causing structures.

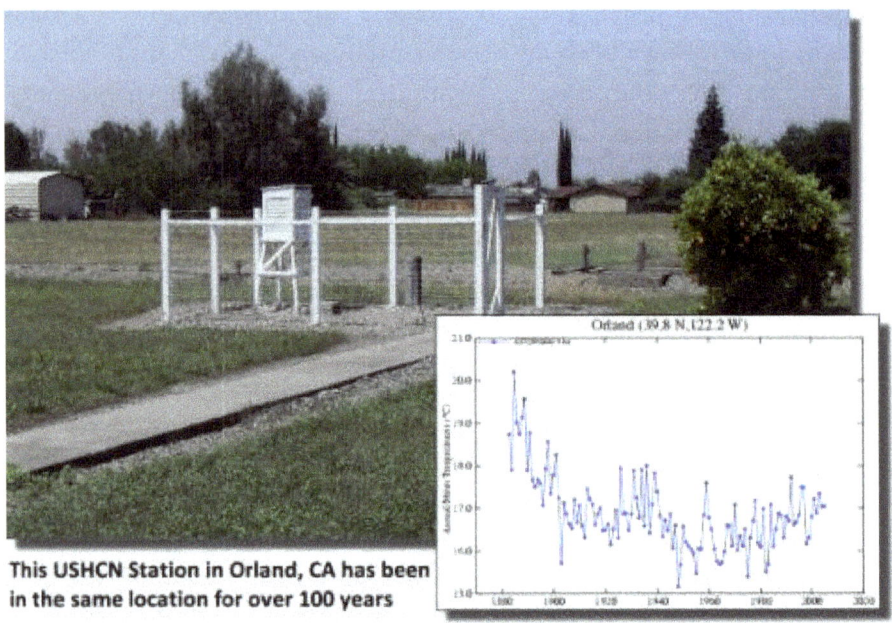

This USHCN Station in Orland, CA has been in the same location for over 100 years

Graph is from https://data.giss.nasa.gov/gistemp/

In 2005, the U.S. launched a "pristine" climate network called USCRN to accurately measure climate variables for the first time from surface stations. These National Oceanic and Atmospheric Administration (NOAA) stations have instruments that are carefully calibrated and maintained at rural sites across the U.S. These are lumped together into one reporting entity which gives us the following graph titled, "Average Temperature Anomaly." This graph doesn't show increases in temperature that the propagandists are purporting. Rather, these readings from the nation's best climate stations actually indicate that the temperature from the surface stations has remained fairly consistent, although seasonally variable, and may even indicate colder average temperatures, or at least much lower "high temperatures" in the summer months.

Chapter Ten: Climate Fake News

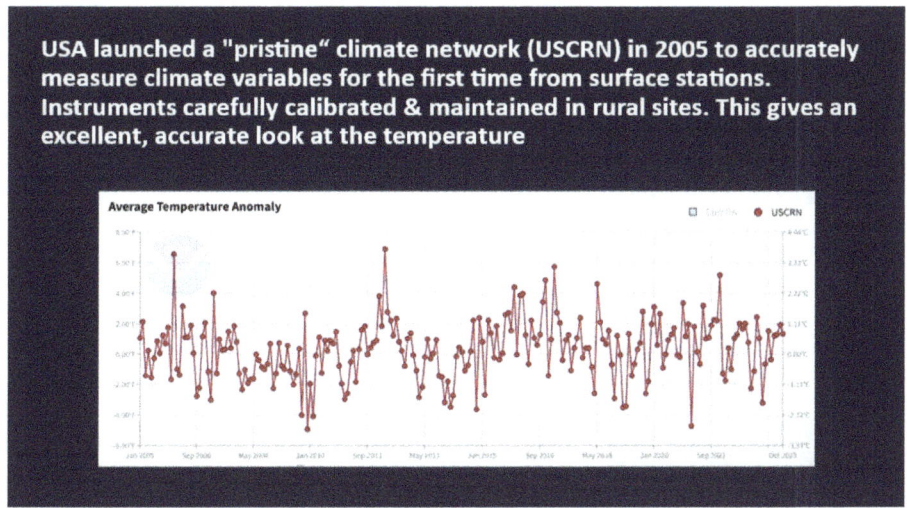

Adjustments to temperature data are necessary to account for changes in station location, instrumentation, or observation practices. However, NOAA does not make adjustments that adequately address factors like Urban Heat Islands (UHI) or station siting issues, leading to a false heat bias in temperature records.

The practice of data estimation, where missing or incomplete data is filled in using statistical methods, is also contentious. It introduces uncertainties and relies on assumptions that doesn't reflect actual conditions. Naturally, these assumptions support the climate change narrative. The numbers from "data estimation" are virtually "pretend" temperatures.

How often does this practice of "data estimation" (read: falsifying information) occur? Sometimes these made-up estimations comprise as much as 75% of all station data[57]. A third of USHCN weather stations have been decommissioned, yet NOAA still makes up their "phantom" temperature data[58].

---

[57] https://electroverse.info/ushcn-weather-stations-decommissioned-yet-noaa-still-uses-their-data/
[58] https://electroverse.info/ushcn-weather-stations-decommissioned-yet-noaa-still-uses-their-data/

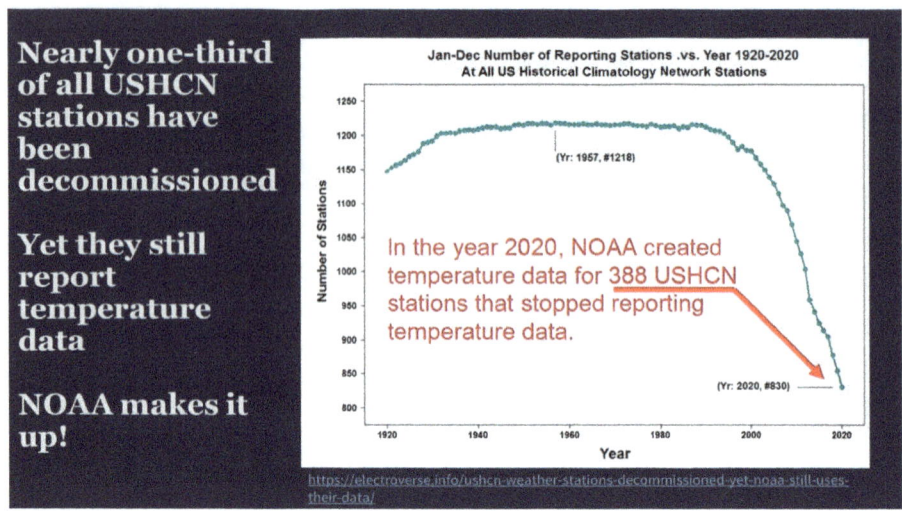

Graph and Description Courtesy of www.youtube.com/watch?v=hs-K_tadvel

The next graph illustrates alternative temperature records between 1979 and 2024 as reported by the NASA satellite record and analyzed by Dr. Roy Spencer at www.drroyspencer.com. Per the NASA satellite record, El Niño storms have caused steep temperature increases since the 1980s. However, satellites don't actually measure temperatures; they measure light energy, running those variables through a computer to create a temperature.

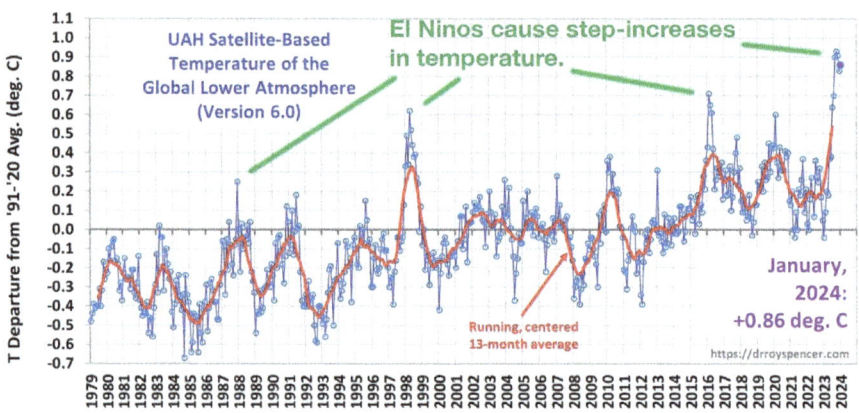

"Temperature.Global at www.temperature.global, calculates the current global temperature of the Earth using about 60,000 stations around the world. It utilizes unadjusted surface temperatures for accuracy. Current temperature is determined by analyzing the 12-month average mean surface temperature and comparing it against the 30-year mean. The site is continuously updated with new observations entered every minute[59]."

Per the real-time surface station record maintained by Temperature Global in this next chart below, there still has been no global warming since the last El Niño in 2015-2016. This is clearly shown by the "Global Temperature" graph below which is the most believable global temperature to which we have access.

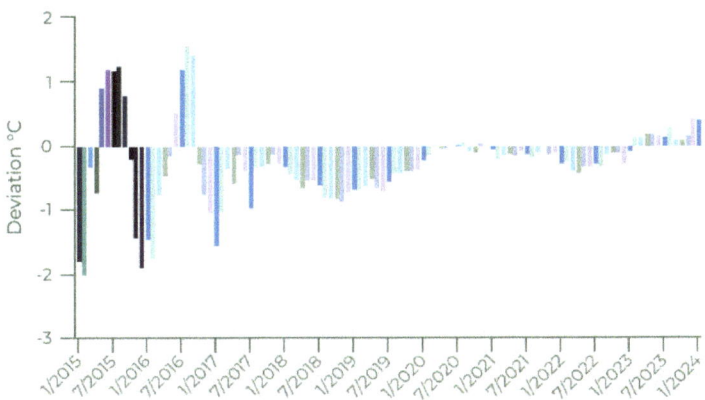

Average (Jan 2015-Jan 2024): -0.258°C
Source: Temperature.Global
Data: NOAA global METARs 2015-current
NDBC global buoy reports 2015-current
MADIS Mesonet Data, NOAA OMOs
https://temperature.global

Graph courtesy of https://temperature.global/

How do these two analyses coexist?

---

[59] https://temperature.global/

Obviously, one of them is false. Or they are both measuring different things. And both are not as precise as they make them out to be.

I'll bet you can take a stab at which one that might be...

However, as stated throughout this book, it doesn't matter because the rise in global temperatures is a part of our Earth's natural cycle of fluctuating weather, climate and temperatures. What's more, hundreds of the world's best scientists agree that it's not at any sort of level to harm Earths' inhabitants, including humans, ecosystems, animals, or any other organism that we can identify. (See Chapter 8.)

While we are on this subject of data stations and accuracy, the next image illustrates where data stations are located globally. As you can see, there are thousands of miles between many stations in the oceans and on most continents. Do you think there are enough data stations for an accurate global temperature?

Locations for global weather stations. From these stations global average temperature is estimated

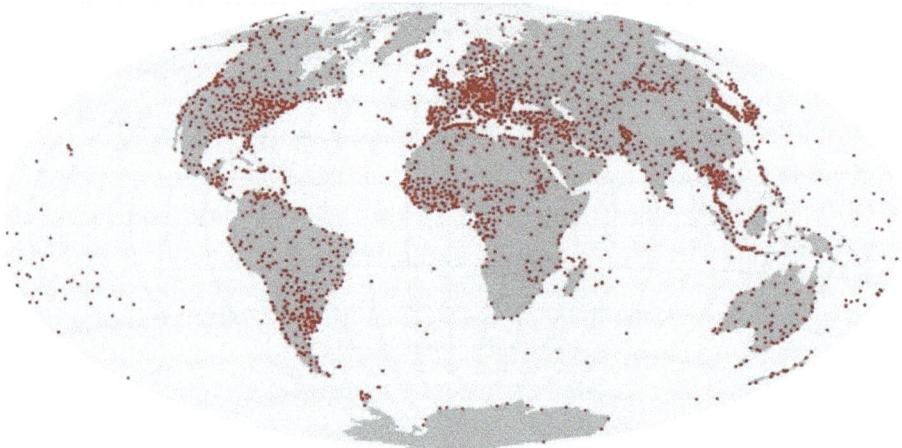

Data Sources:
NOAA Global METARs: Meteorological Aerodrome Reports (METARs) from the National Oceanic and Atmospheric Administration (NOAA) provide surface weather observations from airports worldwide.

NOAA One-Minute Observations (OMOs): NOAA's One-Minute Observations offer detailed, high-resolution weather data collected at various locations.

NBDC Global Buoy Reports: Global Buoy Reports sourced from the National Buoy Data Center (NBDC) provide valuable information on sea surface temperatures and other oceanographic parameters.

MADIS Mesonet Data: Data from the Meteorological Assimilation Data Ingest System (MADIS) includes observations from various mesoscale weather networks, contributing to a comprehensive understanding of surface weather conditions.

Take note that there are almost *no stations* on Antarctica. Would we report global cooling if there were? Would we report global warming if we opened more in the warm or hot equator areas? What about the fact that one-third of the station data is made up in the U.S.?

It is not possible to get a real global temperature, and the data we have is so unreliable, it's laughable.

In 1885, there was almost no climate station coverage outside of Western Europe and the Eastern United States. That said, do you feel confident that we have any remotely accurate idea of how much temperatures have either warmed or cooled? Truly, that is a preposterous notion.

GISS Surface Temperature Analysis (v4)

Station Data, Mercator Map Selector

In our analysis, we can only use stations with reasonably long, consistently measured time records. This is a subset of the full list of stations. That subset of list of stations that contribute to the final products may slightly change with each update, as the number of stations that get dropped due to the shortness of their temperature record may decrease when new data are added. Notice that as part of the homogenization, all stations with less than 20 years of data are discarded (as seen in part (a) of the figure below).

The figures below indicate

    a. the number of stations with record length at least N years as a function of N ,
    b. the number of reporting stations as a function of time,
    c. the percent of hemispheric area located within 1200km of a reporting station.

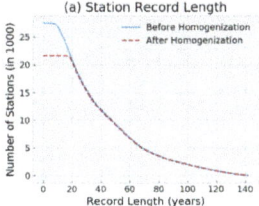
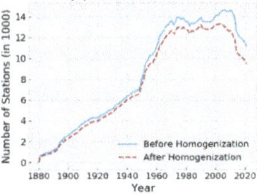
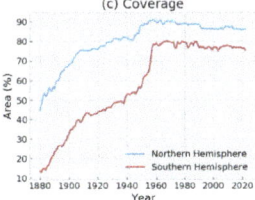

Photo Courtesy of https://data.giss.nasa.gov/gistemp/station_data_v4/

The troposphere is Earth's lowest atmospheric layer, extending from the surface of Earth to about 10 to 15 kilometers (6 to 9 miles). It comprises 75% of the total mass of the Earth's atmosphere and contains 99% of the total mass of water vapor and aerosols. The troposphere is where most weather happens, such as clouds, rain, and storms. It is crucial in regulating the Earth's climate system and it is the layer closest to the surface where humans, animals, and plants live.

According to raw, unadjusted data from Temperature.Global, we do not see an alarming rate of heating; we see a steady climate.

Concerns about the accuracy of temperature data are legitimate and warrant careful consideration in climate research. Transparency in data collection, processing, and reporting is essential for maintaining scientific integrity and addressing public skepticism about climate science. NOAA and NASA are failing.

Tragically, NOAA and NASA cook the temperature data record to fit the man-made climate change narrative rather than reporting the truth.

CHAPTER ELEVEN:
ALTERED DATA & MISINFORMATION

"And you have to wonder: nearly 90% of the thermometers in the U.S. are too close to artificial heat sources. 90%. How much do the climate science team care about the science? 75% of thermometers used in the 1980s have dropped off the official record. All that money, and less instruments to measure with… They adjust the data — sometimes 50 years after it was recorded. Think about that. The 1970s kept warming for the next 30 years."
Joanne Nova

Joanne Nova boasts an impressive academic background, holding a Bachelor of Science degree with exceptional grades from the University of Western Australia. During her studies, she earned prestigious awards such as the FH Faulding and Swan Brewery Prizes. Her academic focus was on Microbiology and Molecular Biology, with research in DNA markers for potential application in Muscular Dystrophy trials. Additionally, she holds a Graduate Certificate in Science Communication from the Australian National University.

As an author, Nova has the consensus narrative surrounding man-made catastrophic climate change through her book *The Skeptics Handbook*. Through her work, she provides alternative perspectives and critical analysis of climate science, fostering discourse and skepticism in the face of prevailing "theory."

During an Anti-Carbon Tax Rally in 2011 she stated:

> *"I used to be a Green. I used to think carbon dioxide mattered. I still worry about falling fish stocks, old growth forest and land erosion, but wind farms won't help the fish, and solar cells won't keep the top soil from blowing away. Real environmental problems are being sidelined by fake ones. Ladies and gentlemen, I thought I was well informed, but I was shocked when I found out what was going on behind the scenes. Everything you may have heard about carbon dioxide can be turned inside out[60]."*

A common misconception is that people who speak out against the climate cult agendas or the Climate Industrial Complex don't care about the environment; but most often, that couldn't be further from the truth. People like myself and Ms. Nova are deeply concerned for the environment, and that is precisely why we want to see resources allocated toward Earth-saving initiatives that actually work, not ineffective corporate greed-fueled tools such as wind towers and solar panels.

When the real data doesn't support the desired narrative, climate change propagandists change it to fit their story, as we've already seen. Nova points out that nearly 90% of thermostats in the United States at the time of her speech were too close to artificial heat sources to present an accurate reading. Warming data is fixed, the data is altered, and a lie is promoted as the truth.

WHY DO NOAA AND NASA LIE?

Unfortunately, lying is a common tactic employed by agencies like NOAA (National Oceanic and Atmospheric Administration) and NASA (National Aeronautics and Space Administration), which receive increased funding from politicians who support the climate change agenda. There are two reasons for this manipulation.

Number one: the agencies need to remain relevant. They don't want their fifteen minutes of fame overshadowed by the truth!

---

[60] https://joannenova.com.au/2011/03/we-are-being-deceived/

Number two: these agencies need to create a problem in order to get money for a solution (to their fake problem.) Since the last manned moon mission, more than 50 years ago in 1972, NOAA and NASA have lost some of their sparkle and sexiness. Naturally, if you're running one of those organizations like a business, you'll fight to have more resources allocated to that business. Climate change serves as a convenient existential threat to justify continued funding NOAA and NASA at higher levels than without it.

Employees at NASA who dissent from the climate change narrative risk repercussions such as termination or demotion. Many retired employees, including astronauts, have spoken out against the narrative, citing inconsistencies in the science. However, their voices are disregarded in favor of the purported consensus.

The issue of temperature adjustments made by NOAA and NASA in their climate datasets is a topic of ongoing debate and scrutiny within the scientific community, but not among the "science is settled" crowd who look the other way when their premises are challenged. This is a form of cognitive dissonance at best, criminal intent at worst. Among the brave naysayers, careers may be stalled or even terminated altogether.

Evidence shows that they adjust past temperatures by making them appear cooler. However, to reflect warming caused by UHI, they should be warming the temperatures and adjust for known biases in the data, such as station location migrations or instrumentation variables.

BLATANT DATA TAMPERING

Let's consider an example in the graph below which shows how raw data for maximum U.S. temperatures compares with adjusted data. The blue line representing unadjusted data illustrates that the average temperature has increased about one degree over the past 120 years. The red adjusted data shows the present having significantly warmed, much more than the blue raw data. The raw data shows that it was warmer in the past than it is now. The warmth of the 1930s and 1940s has been adjusted lower.

Tony Heller is a whistleblower and environmentalist who rides his bike everywhere he goes. Heller holds degrees in both Geology and Electrical Engineering and worked with teams who contributed to the very complex engineering your PC or Mac uses today. He's also worked as a contract software developer for the U.S. government, focusing on climate and weather models.

Heller believes that 97% of claims about global warming are a lie and contends that only 52% of American Meteorological Society members believe that man is a primary contributor to climate change. Heller has downloaded all of the available raw data about temperature changes and has compiled examples that reveal the discrepancies with raw data versus manipulated adjusted data on his blog, realclimatescience.com[61].

The Goddard Institute for Space Studies (GISS) keeps a record maintained by NASA that shows further discrepancies[62].

---

[61] https://realclimatescience.com/?s=raw+data#gsc.tab=0
[62] https://www.giss.nasa.gov/research/features/201501_gistemp/

In the graph below, Dr. James Hansen further illustrates how data has been altered over time to fit the climate change narrative. Hansen, who is widely considered the Godfather of modern climate change hysteria, has faced criticism for his inaccurate predictions and data manipulation, although none of them appear in the mainstream media. This data is greatly influenced by the Urban Heat Island (UHI) Effect.

In the graph below we see how NOAA has adjusted temperatures. They have greatly cooled the past temperatures by as much as a degree and a half and warmed the post 2008 temperatures by more than a degree in some years. These adjustments support the climate change narrative. They have largely erased the cooling I showed you earlier from 1940 to 1979. This isn't science.

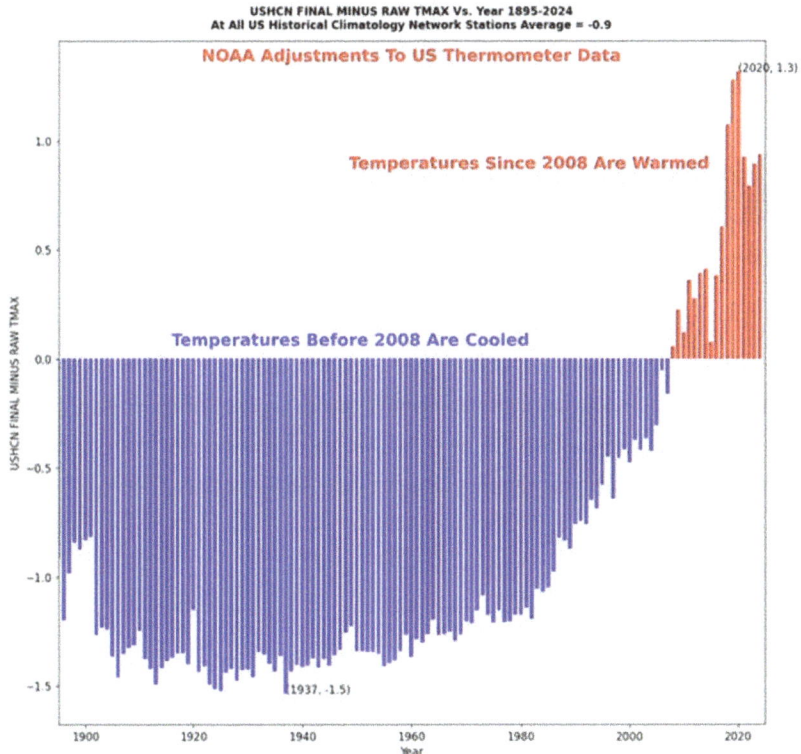

The next image illustrates how NOAA changes the U.S. Temperature Record. The graphs show the same period of time with the same measurements: U.S. temperature in 5-year annual means over 100 years. Of course, the NASA 1999 data differs greatly from the 2019 data. Simply put, NASA changed the chart. They adjusted the climate record to cool the past and warm the present. Rather than a fairly straight and stable line, they adjusted the data to show a more significant warming up.

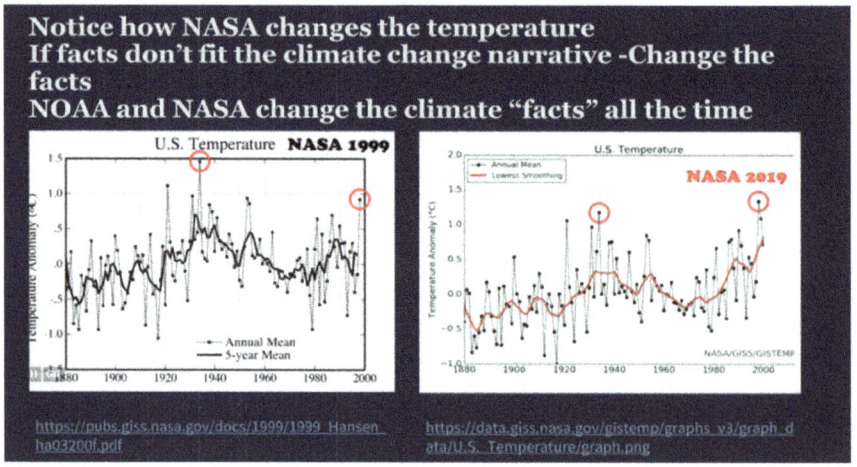

*Both *Original* Photos in this image are courtesy of NASA.Gov*

Further insightful commentary on Nova's blog illustrates the interplay of climate, energy, and politics; a perfect storm of parties invested in shrouding the truth to present false data of global warming. In the following image you'll see how Nova has placed NASA and NOAA graphs side-by-side to illustrate how the same data *changed* when presented in 1980 versus 1987 versus 2007.

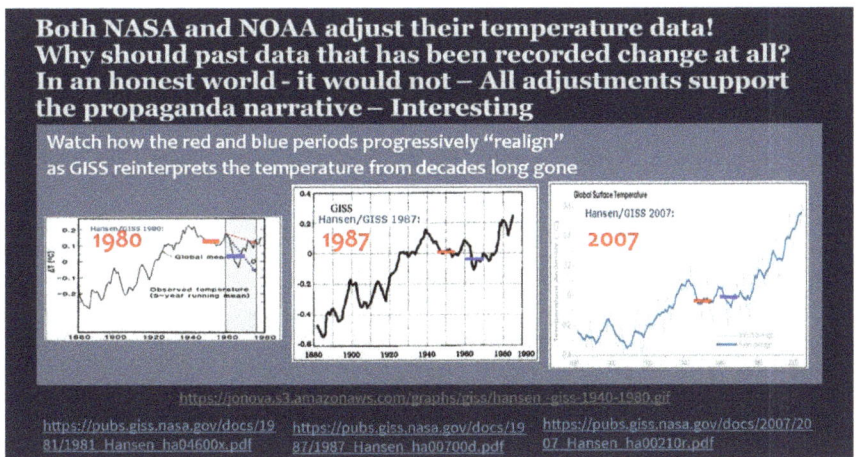

Photo Courtesy of Joanne Nova at www.JoNova.com

The U.S. experienced a cooling trend from the period of about 1940 to 1979. This trend poses a challenge to the narrative of anthropogenic global warming, which asserts that increasing CO_2 levels lead to rising temperatures. To reconcile this discrepancy, adjustments were made to temperature records to erase or minimize the cooling trend.

By eliminating the cooling trend, the data appeared to align more neatly with the global warming theory, providing cleaner support for the narrative. This manipulation of the data allowed proponents of the man-made CO_2-driven climate change narrative to present a more consistent and compelling argument for their cause, regardless of the fact that their argument is only upheld when actual facts are disregarded and data adjusted.

COVERING UP COOLING TRENDS

In 1976, National Geographic also showed a strong cooling trend from 1938-1976. Most importantly, they illustrated that 1976 was no warmer than 1880. During this time period of 1938-1976 there are other instances that illustrate global cooling, not global warming. However, those instances have been largely erased by the propaganda machine, which includes corrupt scientists who alter data by lying.

Because the climate change machine is supported by corporations, politicians, government institutions, and the general (ignorant) public, the scientists conduct these crimes without fear of retribution; their gall is as shocking as their dishonesty.

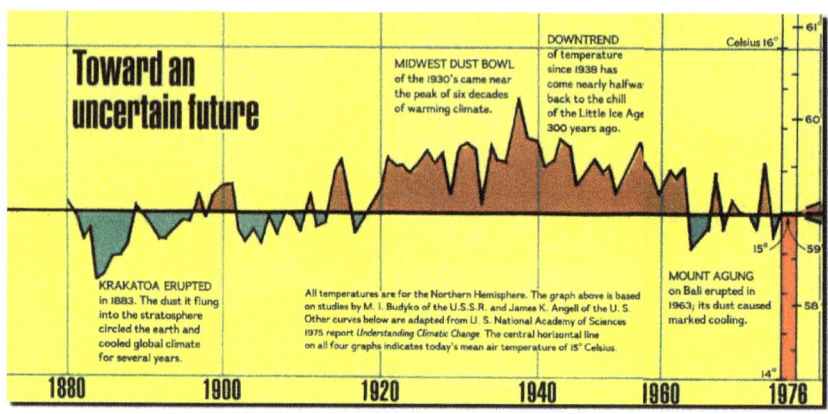

National Geographic Magazine Archive[63]

The newspaper clipping below from the *Santa Cruz Sentinel* in 1989, an article titled "Debate Over Global Warming Heats Up" reports:

> Analysts of warming since 1881 shows most of the increase in global temperature happened before 1919 - before the more recent sharp rise in the amount of carbon dioxide in the atmosphere, said Thomas Karl, of the National Oceanic and Atmospheric Administration's National Climate Data Center in Asheville, N.C. While global climate warmed overall since 1881, it actually cooled from 1921 to 1979, Karl said.

It bears repeating: a scientist from the NOAA stated that since a **rise** in carbon dioxide, global temperatures **cooled,** based upon this 1989 article. It's likely that after publishing this book, this article and many other sources will be summarily erased or censored from the internet, which is one reason I suggest all readers immediately buy physical copies of this book for friends, family members, and your local library.

---

[63] https://archive.nationalgeographic.com/national-geographic/1976-nov/flipbook/614/

> **A-14—Santa Cruz Sentinel — Thursday, Dec. 7, 1989**
>
> # Debate over global warming heats up
>
> to flood low-lying coastal areas.
>
> "The potential consequences are scary as hell," said climate researcher Tim Barnett of the Scripps Institution of Oceanography in La Jolla.
>
> But Barnett and others question whether computer forecasts of climatological doom are valid. That's because so far they have been unable to find evidence to link greenhouse gases to the .5 degree warming observed in the past century.
>
> Analysis of warming since 1881 shows most of the increase in global temperature happened before 1919 — before the more recent sharp rise in the amount of carbon dioxide in the atmosphere, said Thomas Karl, of the National Oceanic and Atmospheric Administration's National Climatic Data Center in Asheville, N.C.
>
> While global climate warmed overall since 1881, it actually cooled from 1921 to 1979, Karl said.
>
> "In spite of all the well-publicized concern about global warming, you must understand that there is still considerable uncertainty among scientific experts about a number of critical factors which determine global warming," NOAA administrator John Knauss said in a statement issued for the geophysics meeting.
>
> Hansen, however, said flatly:

December 7, 1989, Page 14 – Santa Cruz Sentinel at Newspapers.com

CLIMATEGATE

Ten years after this *Santa Cruz Sentinel* article, there was "ClimateGate," another censorship event during which thousands of emails by scientists in the climate field were made public. These scientists discussed back and forth with one another how to manipulate, massage, and alter data to support their global warming climate change narrative. These emails also illustrated that these scientists were involved in creating hysteria.

One email from Tom Wigley to Phil Jones, both with academic email address, "So, if we could reduce the ocean blip by, say, 0.15 degrees, then this would be significant for the global mean-- but we'd still have to explain the land blip... it would be good to remove at least part of the 1940's blip, but we are still left with 'why the blip[64].'" Rather than conducting science

---

[64] https://web.archive.org/web/20130203113349/http://di2.nu/foia/1254108338.txt

and reporting results, these scientists are more concerned with managing a PR campaign in favor of their chosen mission: the cult of climate change.

The "blip" they're referencing in the 1940s during a period of global cooling doesn't support the narrative: it confirms that increased carbon dioxide did NOT affect warming temperatures. ClimateGate is well known in the "Climate Realist" camp. Hackers released about 5,000 emails from several dozen climate scientists, including Michael Mann. It created a stir and when it was covered by the media was deemed "ClimateGate." Professor Phil Jones, the head of British Climate Research stepped down temporarily. There was an investigation, and some of the investigators wanted to see consequences for the scandal.

Ultimately there was only a minor reprimand and the controversy got swept under the rug. There's been a tremendous amount of "spin" to make the event disappear from mainstream news, and the press largely ignored the entire event. Most likely, they were pressured to do so. The spin mainstream media used was that these emails (all 5,000 of them) were taken out of context; it was made to look as if the emails were something innocent; which they weren't.

The blip that is referred to in these emails is that there was a cooling trend from 1940 to 1979 that has largely disappeared in the records[65,66].

---

[65] https://www.forbes.com/sites/jamestaylor/2011/11/23/climategate-2-0-new-e-mails-rock-the-global-warming-debate/?sh=5a89336b27ba
[66] https://www.theguardian.com/environment/2021/oct/10/this-is-a-story-that-needs-to-be-told-bbc-film-tackles-climategate-scandal

Chapter Eleven: Altered Data & Misinformation

```
From: Tom Wigley <wigley@ucar.edu>
To: Phil Jones <p.jones@uea.ac.uk>
Subject: 1940s
Date: Sun, 27 Sep 2009 23:25:38 -0600
Cc: Ben Santer <santer1@llnl.gov>

<x-flowed>
Phil,

Here are some speculations on correcting SSTs to partly
explain the 1940s warming blip.

If you look at the attached plot you will see that the
land also shows the 1940s blip (as I'm sure you know).

So, if we could reduce the ocean blip by, say, 0.15 degC,
then this would be significant for the global mean -- but
we'd still have to explain the land blip.

I've chosen 0.15 here deliberately. This still leaves an
ocean blip, and i think one needs to have some form of
ocean blip to explain the land blip (via either some common
forcing, or ocean forcing land, or vice versa, or all of
these). When you look at other blips, the land blips are
1.5 to 2 times (roughly) the ocean blips -- higher sensitivity
plus thermal inertia effects. My 0.15 adjustment leaves things
consistent with this, so you can see where I am coming from.

Removing ENSO does not affect this.

It would be good to remove at least part of the 1940s blip,
but we are still left with "why the blip".
```

Image Courtesy of: The Wayback Machine

The global warming propagandists, however, continue to defend their belief that carbon dioxide contributes to warming in spite of evidence to the contrary. They claim that increasing sulfur emissions contributed to the cooling trend in the mid-twentieth century. At first glance, it seems reasonable that there would be cooling when you consider the role of sulfur aerosols which would reflect sunlight and thus cool the atmosphere. However, this explanation falls apart when you take into consideration the reality of rising sulfur emissions, particularly from countries like China.

Sulfur emissions have increased substantially in recent years due to China's eight-fold increase in coal usage since 1970, coupled with the absence of widespread clean coal technology. China has lower quality and pollution control measures and their coal is lower quality, making the emissions even more toxic. Nevertheless, the global temperatures still cooled during that period, not warmed. It raises more questions than answers...

RAW DATA ADJUSTMENTS

GISS DATA FABRICATION

NOAA has adjusted the raw temperature data to reflect more warming than is actually happening. In the first two graphs, you'll see images of Annual Mean Temperatures for Reykjavik, Iceland from 2012 and 2013 provided by GISS at NASA. From one year to the next, temperature data is altered so aggressively that the graphs don't even appear to measure the same thing.

Data points from over 100 years of temperature readings are completely altered from one year to the next, obviously to illustrate not the reality of constantly fluctuating temperatures, but to illustrate that we're experiencing a warming trend that is exponentially more significant than actual numbers suggest.

The data has been criminally altered by GISS to show a meteoric rise in temperatures. The actual GISS data illustrates that global warming and global cooling cycles over the past 100 years are a natural, common occurrence. Earth's temperatures fluctuate, they always have, and always will.

In the graphs below, we will see how they are adjusting temperatures around the world. The next graph shows Reykjavik, Iceland. One graph in blue from 2012 shows that 1940 was warmer than today. The graph in red from 2015 shows 1940 as being one of the coldest years and a steady rise in temperature since about 1980.

Chapter Eleven: Altered Data & Misinformation 133

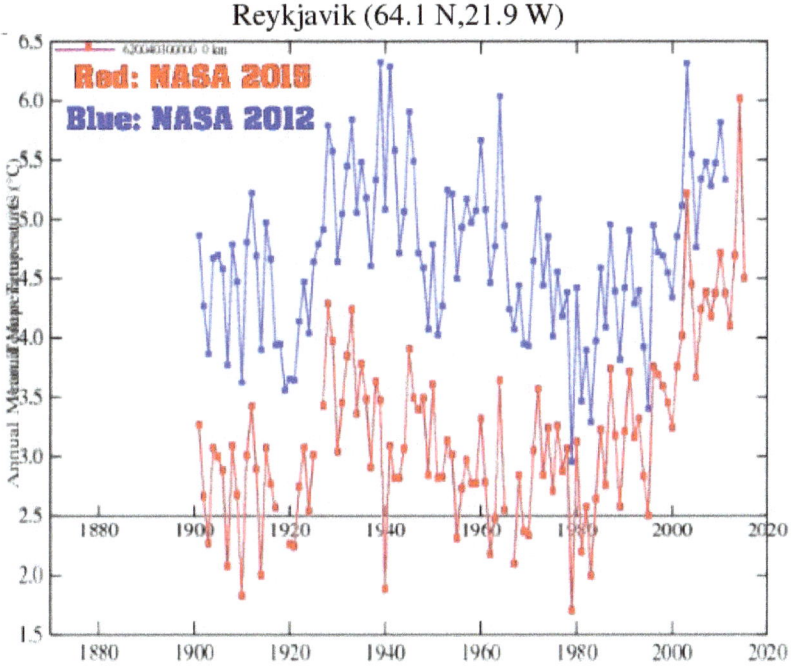

Sources for the two graphs:
2012 version: Data.GISS: GISS Surface Temperature Analysis
https://data.giss.nasa.gov/cgi-bin/gistemp/stdata_show_v2.cgi?id=620040300000&dt=1&ds=1
2015 version: Data.GISS: GISS Surface Temperature Analysis
https://data.giss.nasa.gov/cgi-bin/gistemp/stdata_show_v2.cgi?id=620040300000&dt=1&ds=14

BRITISH DATA FABRICATION

You'll see the same crime in England, here pictured with two graphs illustrating warming in South Wales[67].

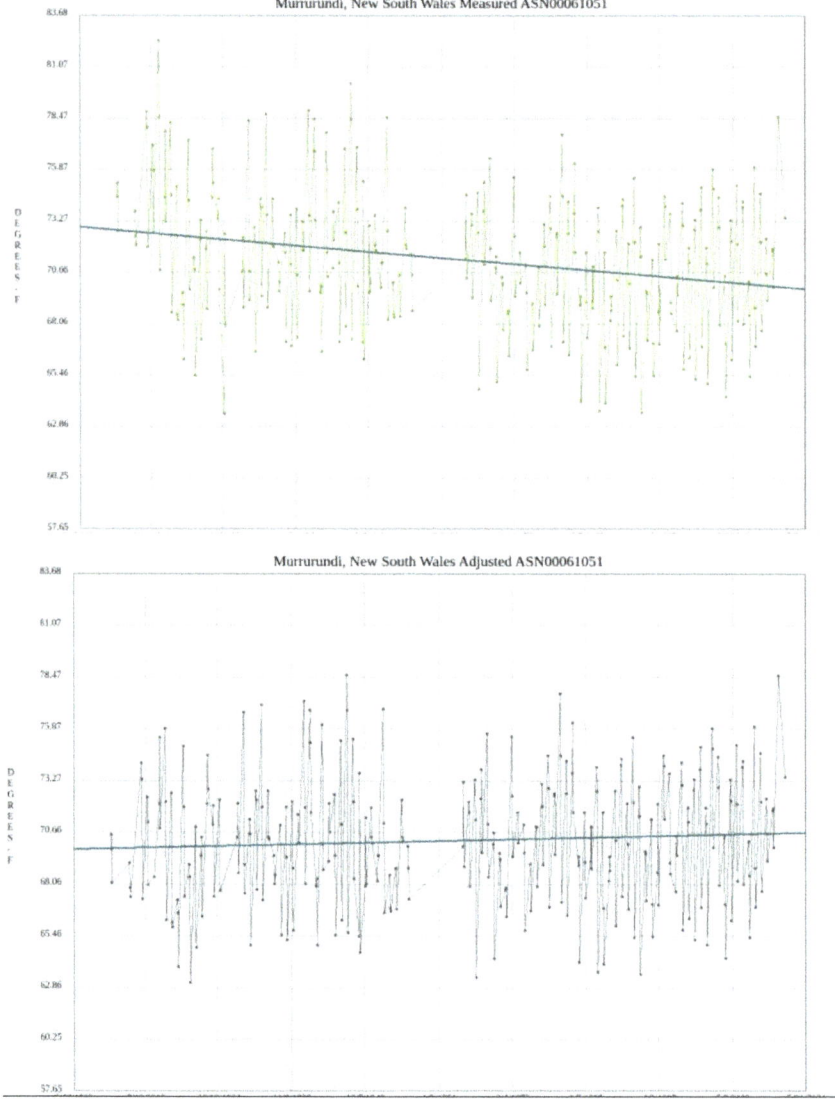

---
[67] https://realclimatescience.com/2015/01/before-they-could-create-the-hockey-stick-they-had-to-get-rid-of-the-post-1940-cooling/#gsc.tab=0

Chapter Eleven: Altered Data & Misinformation   135

ARGENTINIAN DATA FABRICATION

The following graphs illustrate the same data adjustment in Argentina.

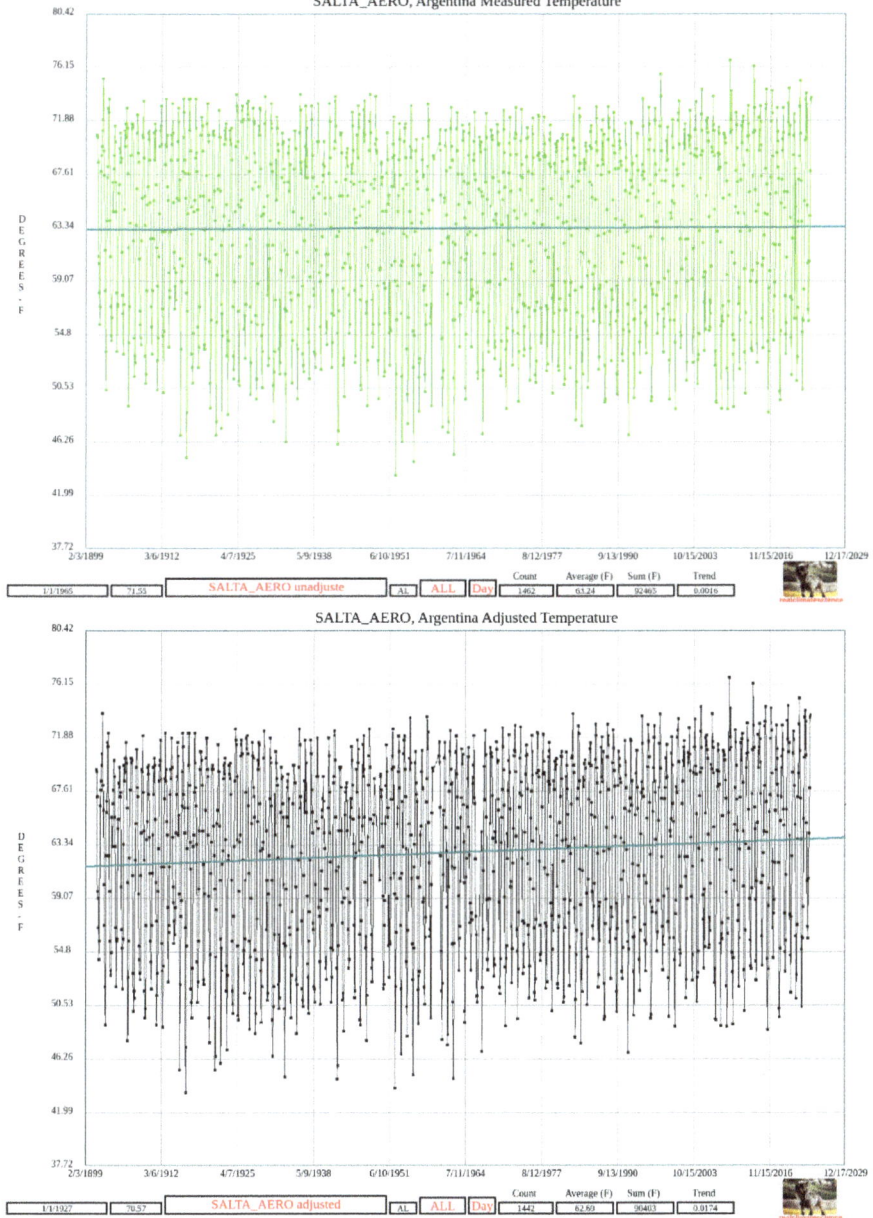

https://realclimate.science/wp-content/uploads/2024/01/SALTA_AERO-adjusted-1.png

URUGUAY DATA FABRICATION

The NOAA adjusted data to show more warming in Uruguay, as well.

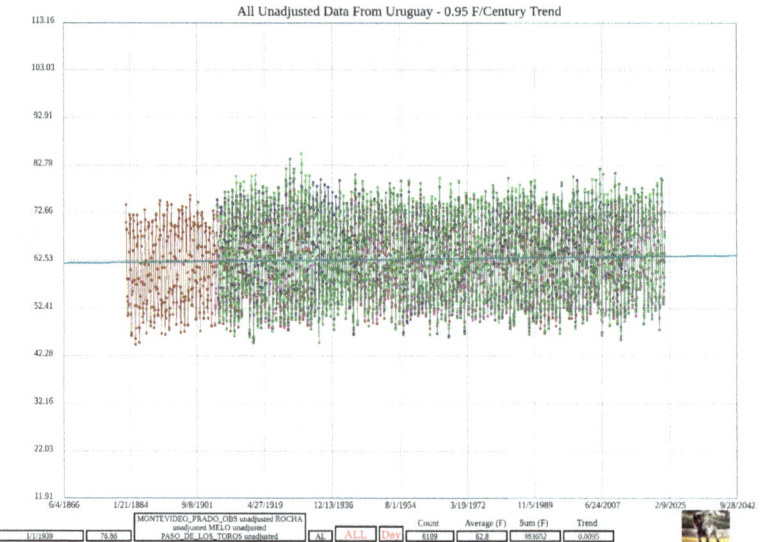

https://realclimate.science/wp-content/uploads/2024/01/MONTEVIDEO_PRADO_OBS-unadjusted_ROCHA-unadjusted_MELO-unadjusted_PASO_DE_LOS_TOROS-unadjusted_MERCEDES-unadjusted.png

https://realclimate.science/wp-content/uploads/2024/01/MONTEVIDEO_PRADO_OBS-adjusted_ROCHA-adjusted_MELO-adjusted_PASO_DE_LOS_TOROS-adjusted_MERCEDES-adjusted.png

Chapter Eleven: Altered Data & Misinformation 137

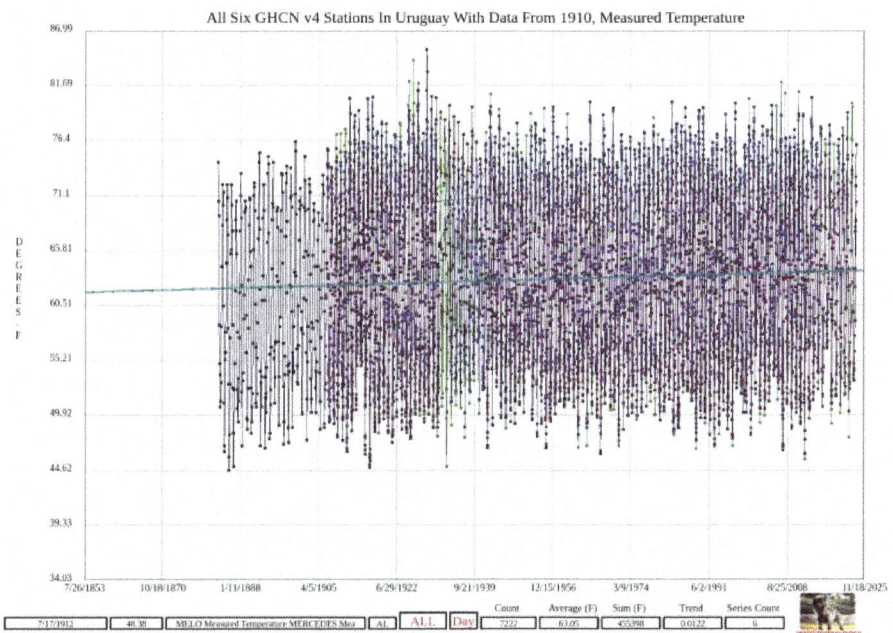

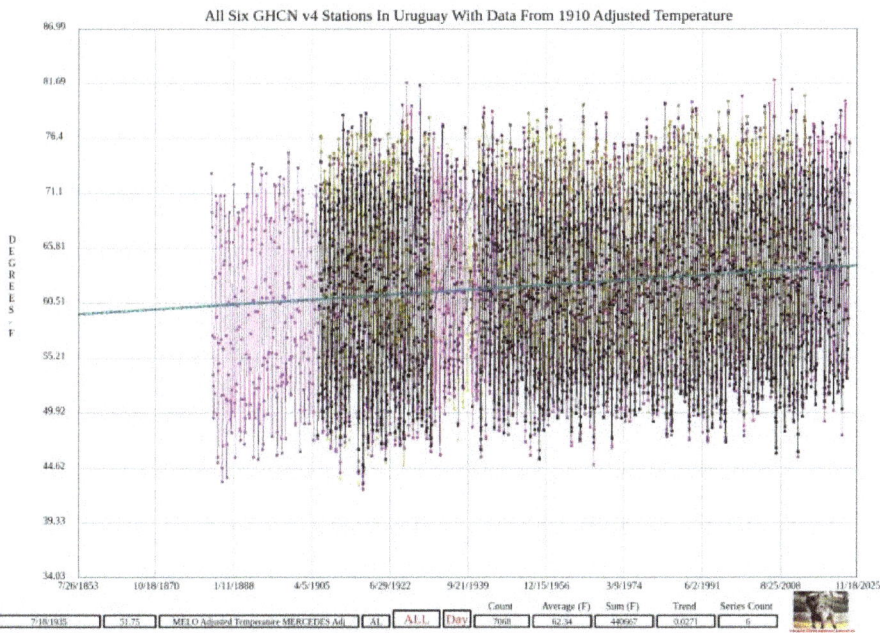

https://realclimatescience.com/2024/01/climate-adjustment-fraud-in-south-america/#gsc.tab=0

VENEZUELA DATA FABRICATION

The same happened in Venezuela.

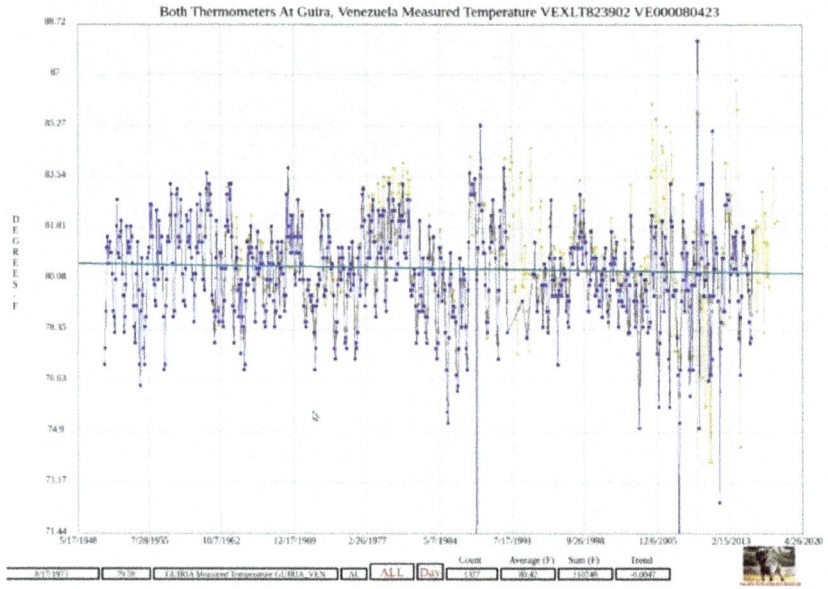

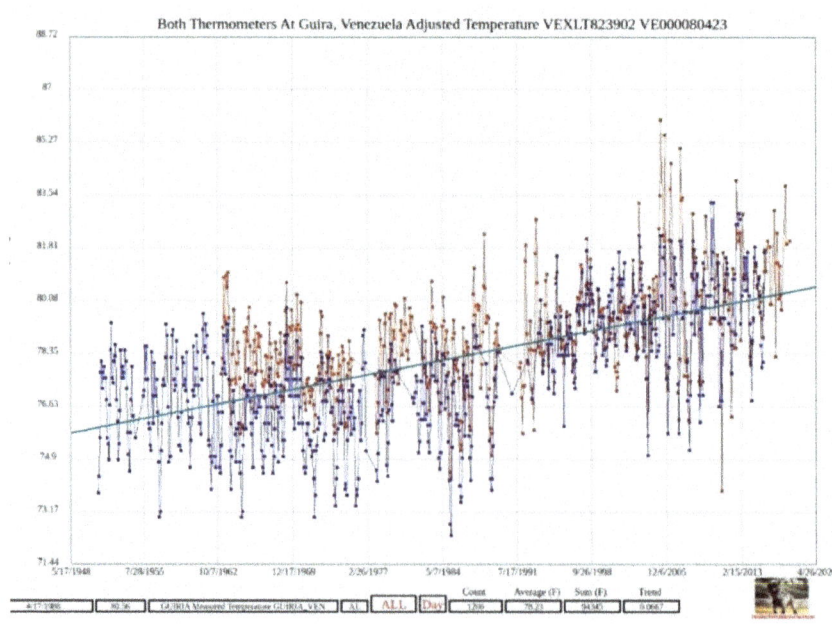

UNITED STATES DATA FABRICATION

Data tampering also occurred in the United States, as the following graphs illustrate with the five-year mean appearing to increase as opposed to the true graph, which illustrates that since around 1925 the Continental U.S. temperature has decreased.

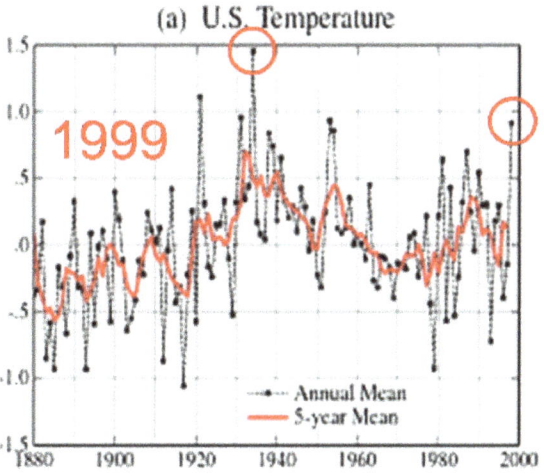

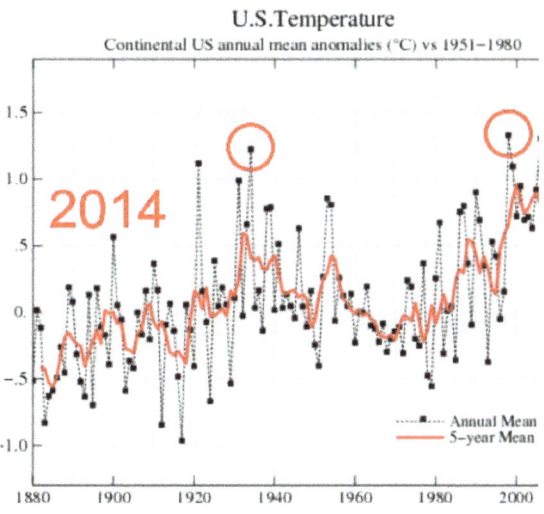

GLOBAL DATA VULNERABILITIES

As we've established, most global weather stations that report data for these graphs are in places that are vulnerable to added heat from synthetic sources; their readings can't be trusted. They should, theoretically, have disclaimers that they are situated in places that increase the temperature readings. The world isn't boiling, the books are being cooked from every angle.

The map below entitled "Land-Only Temperature Departure from Average November 2023 (with respect to a 1991- 2020 base period)" illustrates actual climate stations and measurements. The gray areas mean we do not have any temperature records or measurements in those spaces. As you'll see on this map, there are no measurements in most of Africa, an enormous continent, nor in most of the oceans, which cover 71% of our Earth.

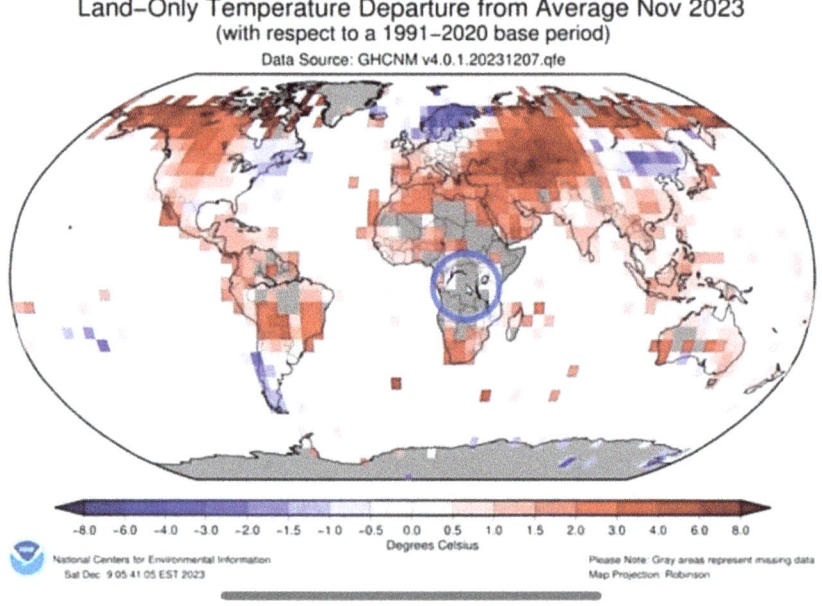

Map Courtesy of the National Centers for Environmental Information Saturday, December 9, 2023

Chapter Eleven: Altered Data & Misinformation 141

AFRICAN TEMPERATURE TAMPERING

The graph below shows the climate stations in Central Africa. As you'll see, there are only a small handful of dots representing almost no data from a massive section of land, including the Congo, Tanzania, Gabon, and Angola, which is illustrated by the colorful graph below the temperature station graph. In short, there are only a few dozen stations to monitor about a third of our second largest continent by both population and size.

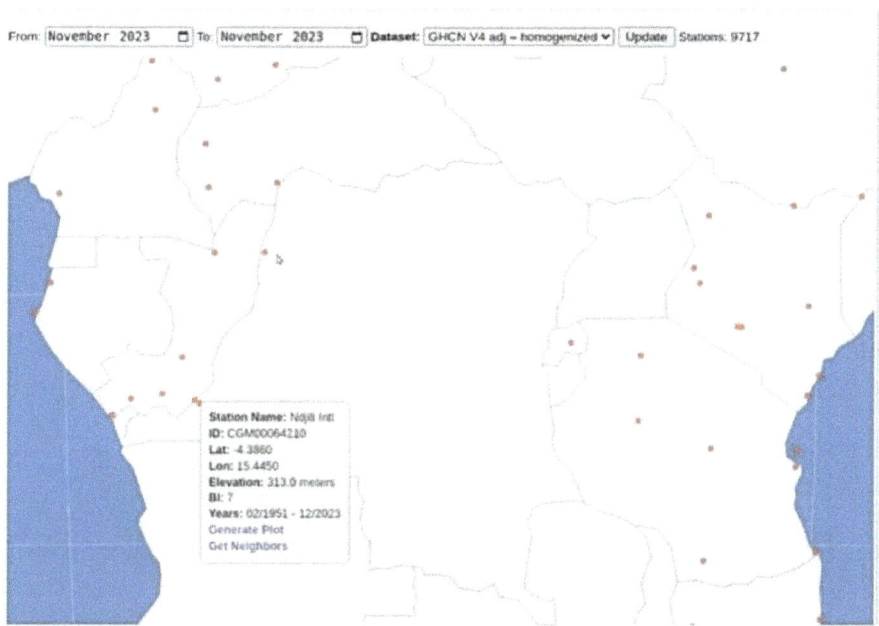

Part Three: Climate Lies

Image Courtesy of Adobe

THE CONGO TEMPERATURE ABSURDITY

The temperatures NOAA tracks for the Democratic Republic of Congo region of Africa derive from a single climate station in Kinshasa, a station located at the airport. Kinshasa has grown tremendously over the last thirty years and their raw data shows the effects of the urban heat island (UHI).

In the graph below entitled "All Measured GHCN V4 Temperature Data from The DRC (Excluding the Kinshasa Airport)," you see all the neighboring temperatures without the airport.

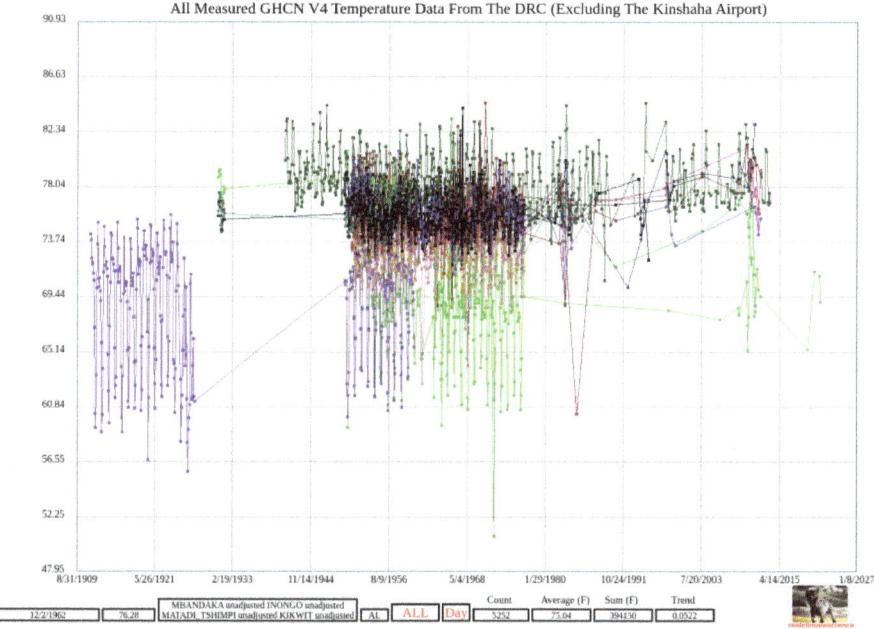

https://realclimate.science/2024/01/24/imaginary-record-heat/mbandaka-unadjusted_inongo-unadjusted_matadi_tshimpi-unadjusted_kikwit-unadjusted_kananga-unadjusted/

Due to UHI as well as population growth, and more plane traffic in and out of the airport, the measured data at the airport shows a distinct warming trend. This seems fairly obvious. This graph doesn't show the temperature for the massive region of the Democratic Republic of Congo, but only for the Kinshasa Airport. Due to the Urban Heat Island Effect and more traffic, it appears that "the Congo" is warming up.

No, it's just one station in a hot location that's warming up. To do actual scientific research, we'd require many more temperature stations all over the Congo in various altitudes, climates, and in a mix of both rural and urban areas.

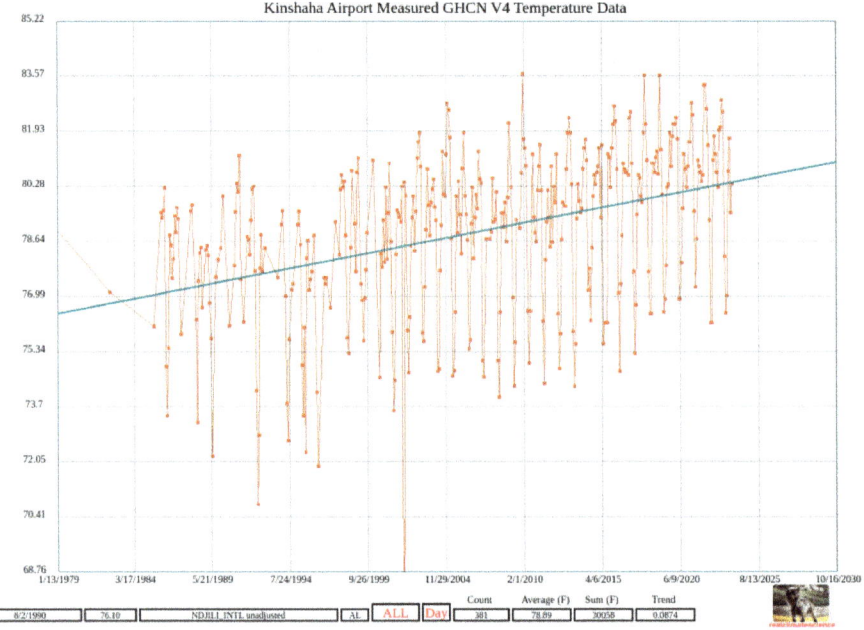

https://realclimate.science/2024/01/24/imaginary-record-heat/ndjili_intl-unadjusted-1/

Chapter Eleven: Altered Data & Misinformation

NOAA FALSIFIED CONGO DATA IN MAP OF THE WORLD

In the map below, you'll see that NOAA clearly falsified data to create this heat map of the world. There are very few temperature stations in central Africa, and many spots that are literally gray with no data coming in whatsoever.

In the graph below from NOAA's National Centers for Environmental Information entitled "Land & Ocean Temperature Percentiles Nov 2023," you'll see a map that's covered with red boxes indicating "Warmer Than Average," "Much Warmer Than Average," and "Record Warmest" temperatures. This map is designed to be compelling; to fear monger viewers into believing that we are experiencing record warm temperatures. Thankfully, this data is largely falsified and taken from stations that aren't taking accurate readings, if they're taking readings anymore, at all.

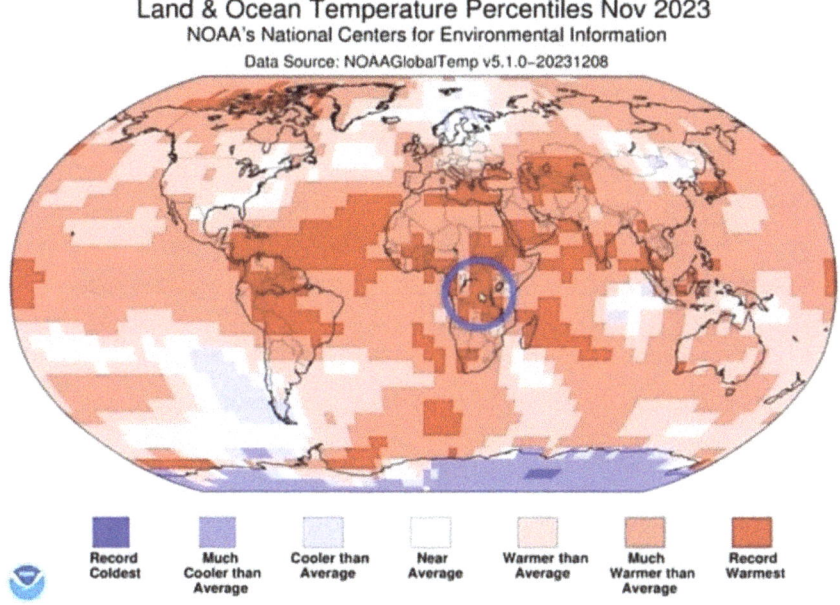

Now, let's return to the map we looked at earlier from the National Centers for Environmental Information that illustrates gray areas in place of the "Record Warmest" areas from the map above. In the map that's been altered by the NOAA's National Centers for Environmental Information, the climate change and global warming narrative seems to be settled in a glance.

The problem is that you're looking at a lie. In actuality, the places that appear to be at record warmth are, in fact, not being tracked at all. Pay close attention to the portion of the map that is circled near Central Africa. Notice how very hot it is in the map above where it shows "Record Warmest" temperatures. We have no information from that area, as the second map clearly illustrates.

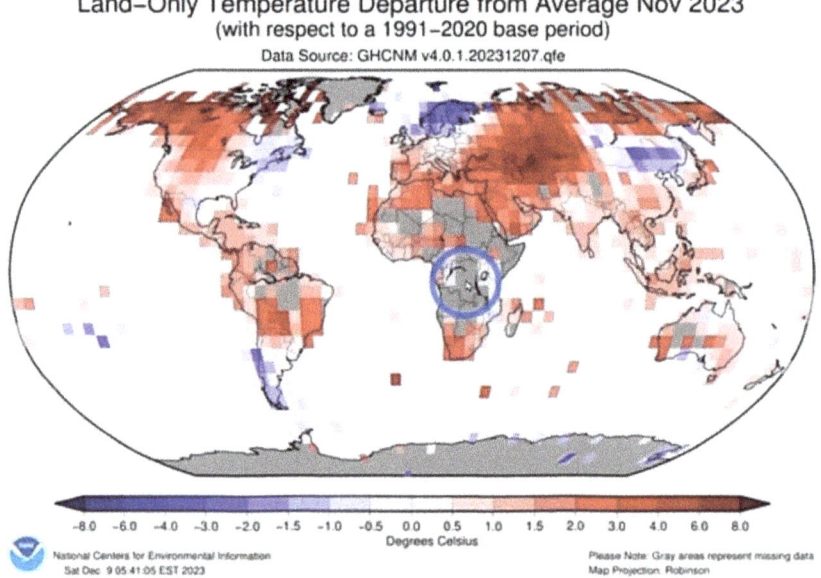

Map Courtesy of the National Centers for Environmental Information
Saturday, December 9, 2023

If you're like me, your blood boils when you see these maps. I've been censored from social media for sharing similar maps and without the intrepid journalists and podcasters who will courageously provide me a stage during this book's tour, billions of people may never hear nor see the truth; although it is so clearly "mapped out" in front of our faces.

NOAA is lying. Politicians are lying. Scientists are falsifying data. Maps and graphs are inaccurate. Folks, let's get together to share the truth about manipulated temperature data.

In Part Four of this book, I'll share with you a story about how energy and food prices could soon skyrocket in order to fall in line with the wishes of the cult of climate change, while politicians are justifying policies and regulations that harm our freedom and prosperity.

My hope is that by the end of this book, you'll not just feel outraged at the blatant lies you've been fed, but that you will also feel empowered with some specific ways YOU can help the world overcome the cult of climate change to save our sovereignty for generations to come.

CHAPTER TWELVE:
IT'S NOT AS HOT AS IT USED TO BE

"The growth of cities makes it hotter, but that was true back in the 1930's, too. Big cities were hotter than the surrounding countryside because you concentrate the traffic, and you concentrate the home heating. And you modify the surface, you pave a lot of it...It has gone up since the early 1800s, before the Industrial Revolution, because we're coming out of the Little Ice Age, not because we're putting more carbon dioxide into the air. You can go outside and spit and have the same effect as doubling carbon dioxide."
Dr. Reid Bryson, the "Father of Climatology"

In this chapter, we'll explore the concept of "hotter days" and rising temperatures next to another set of maps, graphs, and, of course, propaganda.

Dr. Reid Bryson was a prominent American atmospheric scientist and geoscientist, who was regarded as the "father of scientific climatology." Bryson was known for his groundbreaking work in climate science, particularly in the field of paleoclimatology, which involves studying past climates using geological and other proxy data.

Bryson was one of the first scientists to propose the concept of "climate forcing," which refers to factors that can cause changes in the Earth's climate, such as variations in solar radiation, volcanic activity, and greenhouse gas concentrations. Bryson served as a professor at the University of Wisconsin–Madison and founded the Center for Climatic Research at the same institution. He also served as director of the Institute for Environmental Studies at the University of Wisconsin–Madison.

In 1977 Bryson along with Thomas A. Murray published a book called, *Climates of Hunger*, the opening page of which states,

> "Climate is changing. Parts of our world have been cooling. Rain belts and food-growing areas have shifted. People are starving. And we have been too slow to realize what is happening and why. In recent years, world climate changes have drawn more attention than at any other time in history. What we once called "crazy weather," just a few years ago, is now beginning to be seen as part of a logical and, in part, predictable pattern, an awesome natural force that we must deal with if man is to avoid disaster of unprecedented proportions.[68]"

> Climate is changing. Parts of our world have been cooling. Rain belts and food-growing areas have shifted. People are starving. And we have been too slow to realize what is happening and why.
>
> In recent years, world climate changes have drawn more attention than at any other time in history. What we once called "crazy weather," just a few years ago, is now beginning to be seen as part of a logical and, in part, predictable pattern, an awesome natural force that we must deal with if man is to avoid disaster of unprecedented proportions.
>
> Along with drought in some places and floods in others, both caused by changing wind patterns, average temperatures of the Northern Hemisphere have been falling. The old-fashioned winters our grandfathers spoke of might be returning. In England, the growing season has already been cut by as much as two weeks. The selection of food crop varieties in both North America and Europe is in for sharp reappraisal, in view of the shrinking frost-free agricultural season and other climatic changes.
>
> Climate has always had profound effects upon human history, helping both to build and to destroy great civilizations. Until now, we have not had the knowledge to react intelligently to the signs of shifting climate. Today, even though we remain essentially powerless to affect climate purposefully, we are ready to recognize the signs of change and we are somewhat better able to predict the effects of those changes.
>
> This book will help. Here, climatologist Reid A. Bryson and science writer Thomas J. Murray present a broad view of climatic change, examining the past in order to view the future.
>
> The prospects are not bright. Bryson, whom *Fortune* magazine called "the most outspoken perceiver of climatological danger signals" in the United States, says that world temperatures since the sixteenth century have been significantly cooler than those of the first half of the present century. Temperatures now seem to be falling, and many of the weather irregularities we have experienced in recent years are, in great part, an expression of this broad reversal.

Excerpt from the introduction of "Climates of Hunter" by Reid Bryson and Thomas A. Murray

---

[68] https://openresearch-repository.anu.edu.au/bitstream/1885/114865/2/b11619600.pdf

Today, Bryson provides us with some of the most poignant confirmation of the "Urban Heat Island" effect as well as other admittances that in some parts of the United States and the world, there is no warming at all. He says,

> *"For example, in Wisconsin in the last 100 years the biggest heating has been around Madison, Milwaukee and in the Southeast, where the cities are. There was a slight change in the Green Bay area. The rest of the state shows no warming at all...*

The narrative that there are "more heat waves" and "hotter days than ever before" is plainly not supported by historical data. Claims of increasing heat wave frequency, heat wave duration, and heat wave intensity are not supported by evidence.

In the graph below, you'll see trends in multi-day extreme heat events across the U.S. from 1895 to 2020. Of course, there's a chance this graph will be altered in order to fit political narratives or censored in due time. While we have access to it, let's note the peak of heat waves in the United States. They were all the way back in the 1930s, nearly *one hundred years ago*. This obviously contradicts the notion of a current upward trend in heat waves.

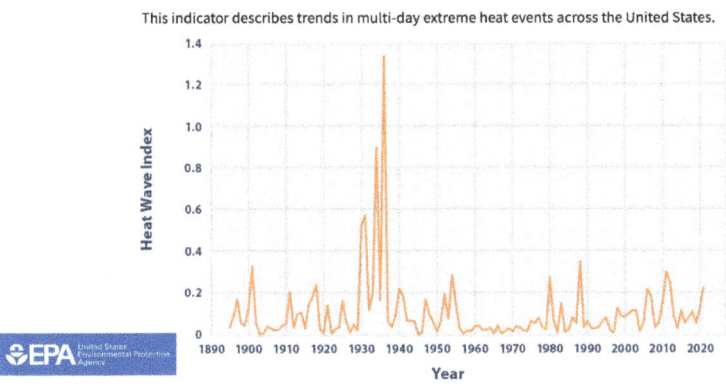

https://www.epa.gov/climate-indicators/climate-change-indicators-heat-waves
Accessed March 11, 2024

This graph is from a different EPA webpage and in the report on EPA.gov, you will see the following set of graphs indicating heat waves by number of heat waves per year or "frequency," length of heat waves in days, length of annual heat wave season, and average temperature above the local threshold during the heatwave[69].

Upon first glance, you might notice that the heat wave data appears to be increasing based on frequency, duration, season, and intensity. However, data from NOAA and other sources show that there is no significant upward trend in the number of hot days. No more all-time record highs have been set in recent years than statistical randomness would be suggested in a non-warming climate.

Furthermore, Urban Heat Island Effect (UHI), where cities experience higher temperatures due to factors like increased concrete and lack of vegetation, plays a significant role in local temperature increases. As we've seen, as cities (or airports) expand, they become warmer, which contributes to rising measured temperatures. Again, cities can be 5° to 7° F (2.8°C to 3.9°C) warmer than the surrounding countryside. In the four graphs below, you don't see the increase in UHI effect, which makes these graphs misleading.

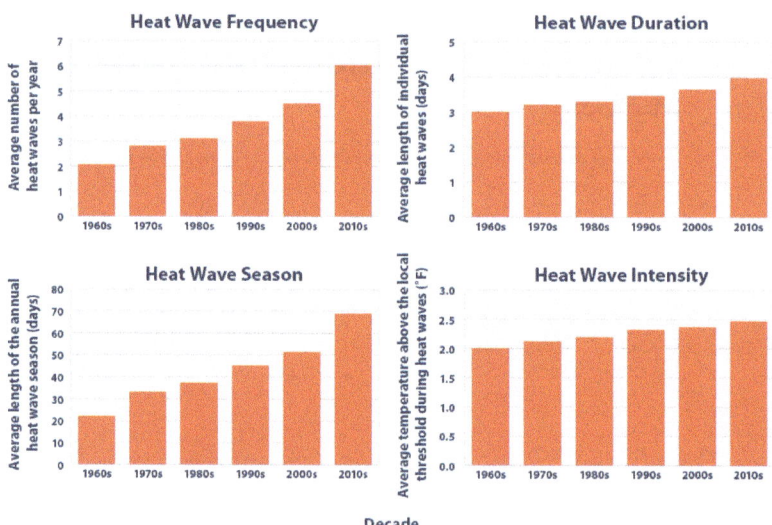

https://www.epa.gov/climate-indicators/climate-change-indicators-heat-waves
Accessed 12/26/22

---

[69] https://www.epa.gov/climate-indicators/climate-change-indicators-heat-waves

The EPA maps below show heat waves across the U.S. based upon the same information as above: frequency, duration, season, and intensity. The EPA maps illustrate heat wave increases precisely where there are large cities, starting with Seattle and Portland, and traveling as far as New York City and Miami on the East Coast.

These EPA maps don't show that we're experiencing heat waves or increases in frequency, duration, season, nor intensity. Rather, this map illustrates the effect of Urban Heat Islands perfectly. Certainly, the increases in heat wave factors have nothing to do with carbon dioxide, and everything to do with the fact that these temperatures are being recorded in cities that are, obviously, heating up due to population growth and industrialization.

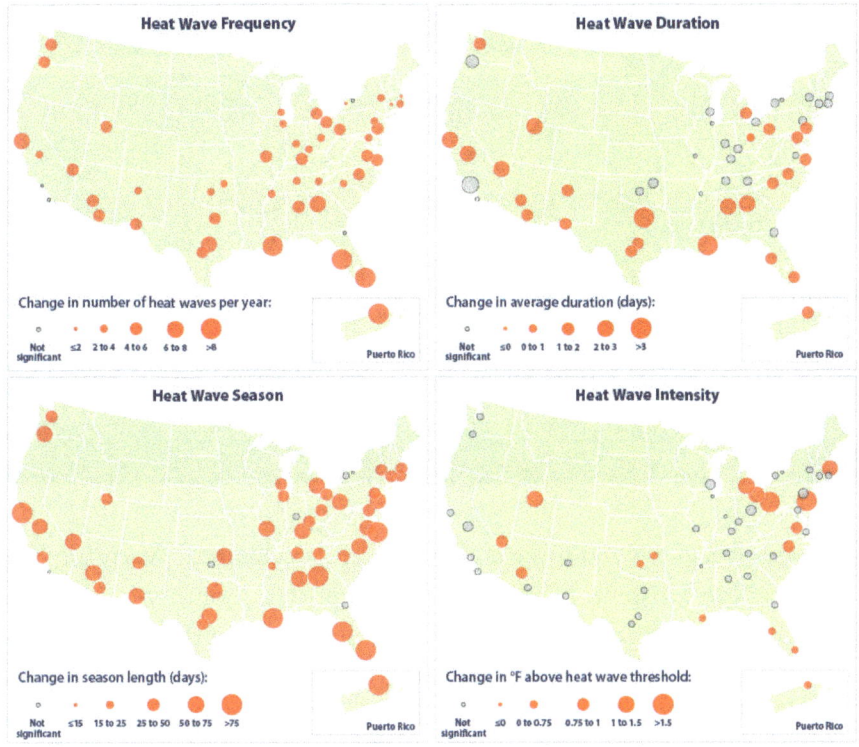

https://www.epa.gov/climate-indicators/climate-change-indicators-heat-waves
Accessed 12/26/22

RECORD HEAT IN 1936

In the U.S., there are 1,218 USHCN temperature gauge stations. In 1936, 245 of them set their all-time record maximum temperatures.

In the image below you'll see that on July 7, 1936, twenty-four states were over 100 degrees Fahrenheit. North Dakota, Indiana, Illinois and Kentucky were above 110 degrees Fahrenheit, and the average afternoon temperature was above 90 degrees Fahrenheit. This was nearly a hundred years ago. There was industrialization during this time, but the population was vastly smaller than it is today.

In the graph below, you'll see a chart "U.S. 'Accelerated' Warming Since 1895: It's Non-Existent. NOAA/NCDC reported monthly temperatures and CO_2 levels." The author of this chart points out that the red line, which shows the average temperature, appears to be a straight line with very little variation. The dark dots that show an upward trajectory show increasing CO_2. What this evidence shows is that we have a "less cold" world, not a hotter world.

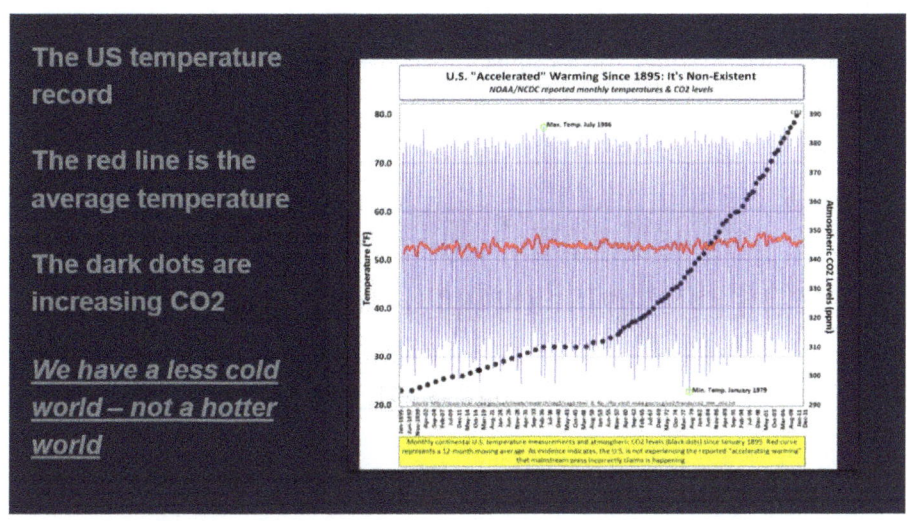

https://www.ncei.noaa.gov/access/monitoring/climate-at-a-glance/national/time-series/110/tavg/1/0/1895-2024?base_prd=true&begbaseyear=1895&endbaseyear=2024

From the Lansing State Journal on Monday, December 12, 1988, the NASA scientist often known as the "Father of Modern Climate Change Hysteria," James Hansen wrote:

> *"If you liked last summer's record temperatures, you're going to love the 1990's, says James Hansen, the NASA scientist who, during congressional hearings on the Midwestern drought, linked greenhouse warming to the heat wave...*
>
> *Although many scientists argue that the dry, hot summer of '88 was not caused by greenhouse warming, it's hard to find a climate expert who will claim that greenhouse effect is not on its way.*
>
> *... Washington, D.C., for instance, would go from its current 35 days a year over 90 degrees to 85 days a year. The level of the ocean will rise anywhere from one to six feet."*

Hanson made a false prediction in 1988 when he stated that the oceans would rise or that we would have 85 days a year of over 90-degree temperatures in D.C. That isn't all Hansen has been wrong about, but that doesn't stop him from continuing to make false predictions. The media continues to interview this unreliable source, and because he gets "clicks," he never gets asked to explain past inaccuracies.

Prepare for long, hot summers

By EDWARD STILES
Gannett News Service

If you liked last summer's record temperatures, you're going to love the 1990s, says James Hansen, the NASA scientist who, during congressional hearings on the Midwestern drought, linked greenhouse warming to the heat wave.

Last summer was a preview of the average summer 10 years from now, and the hottest summers during the '90s will be even hotter and drier than the one we just struggled through, he says.

Although many scientists argue that the dry, hot summer of '88 was not caused by greenhouse warming, it's hard to find a climate expert who will claim that the greenhouse effect is not on its way.

When Hansen, head of the Goddard Institute for Space Studies, spoke recently to researchers at the University of Arizona Lunar and Planetary Laboratory, he ticked off several unpleasant changes in the weather most scientists agree probably will occur during the next 50 to 60 years:

■ If we do nothing to cut down on pumping carbon dioxide into the atmosphere, temperatures in 2050 will be 6 to 7 degrees higher than they are today. Washington, D.C., for instance, would go from its current 35 days a year over 90 degrees to 85 days a year.

■ The level of the ocean will rise anywhere from one to six feet.

■ The frequency and severity of storms would increase. If the amount of carbon dioxide in the atmosphere doubles — the worst-case scenario between now and 2050 — the maximum strength of hurricanes may increase by 50 percent, Hansen says.

While a few degrees warmer or cooler may not seem like much, such a change can result in huge differences in climate. Hansen notes that during the last ice age the earth was only about 9 or 10 degrees cooler on average than it is now.

Lansing State Journal, Monday December 12, 1988

Today in 2024, we've seen no such upward trend in the Washington D.C. area of increased heat waves nor temperatures. When Hansen testified before the Senate, he happened to be testifying during a warm summer. It is reported that they also opened the windows and shut off the air conditioning the day before the hearing so that it would be very hot in the hearing room to add to the hysteria of climate cultists. Today, I doubt many politicians would require such manipulation to buy into climate hysteria because society seems to have become increasingly willing to buy into media propaganda.

I presume that the nefarious manipulation by Hansen derives from his desire to raise more money for NASA. He was able to get politicians such as Al Gore to align with him and his cause. In the graph below, you'll see data from Lincoln, Virginia, showing days above 90 degrees Fahrenheit.

However, the graphs below also show that the overall number of days above 90 degrees Fahrenheit has, in fact, decreased since 1911 when they peaked[70].

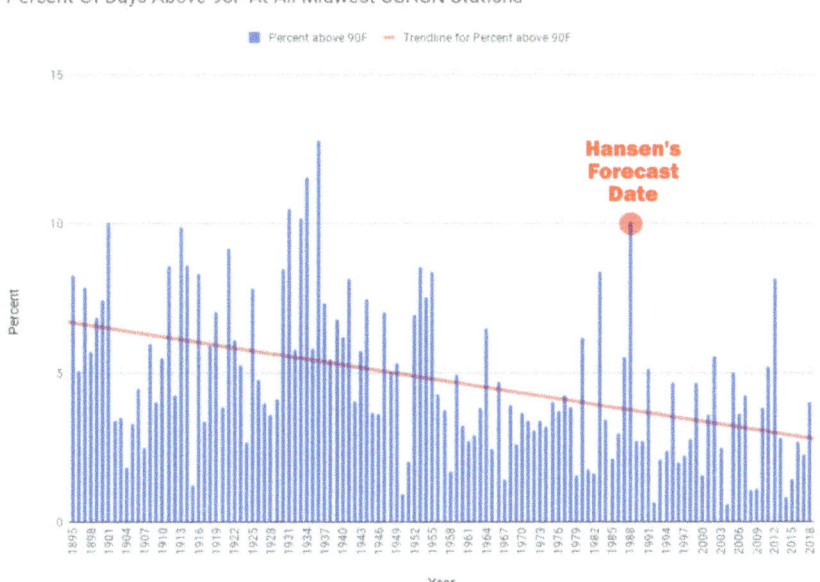

Photo Courtesy of RealClimateScience.com

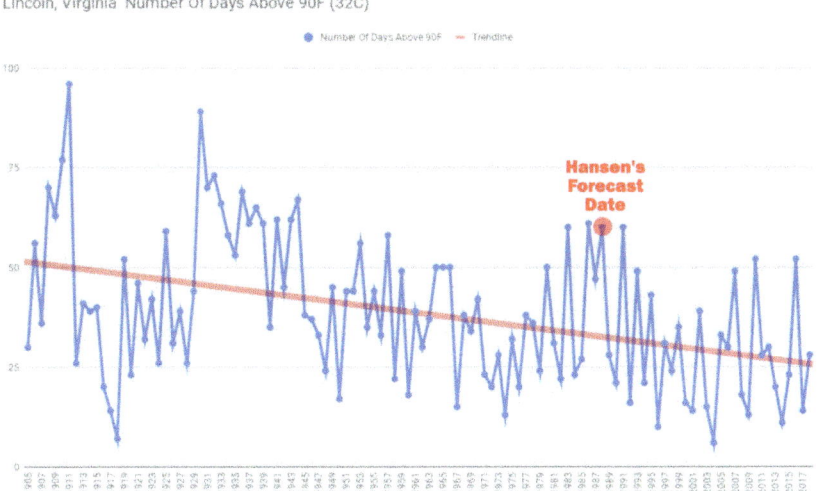

Photo Courtesy of RealClimateScience.com

---

[70] https://realclimatescience.com/2019/05/hansen-got-everything-wrong-alarmists-claim-victory/#gsc.tab=0

https://www.youtube.com/watch?v=CyI2Hp8YxjM&t=87s
https://realclimatescience.com/2022/08/the-climate-emergency-of-1955/#gsc.tab=0

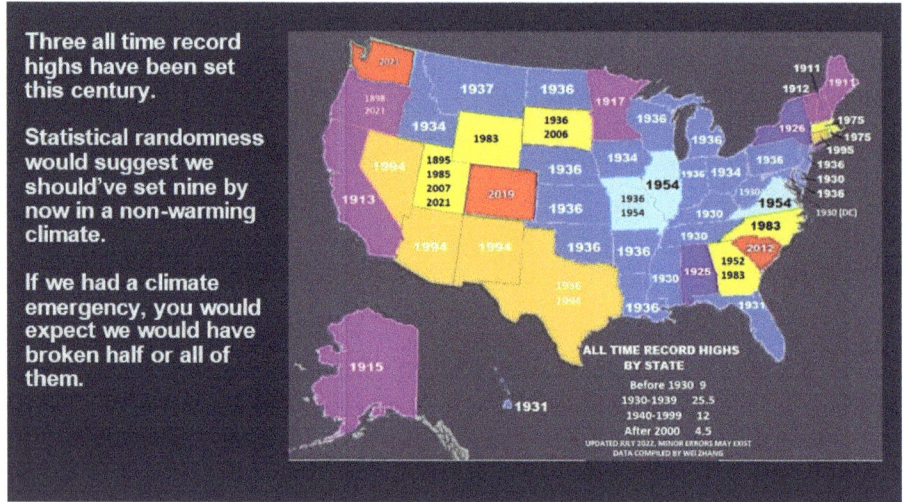

https://www.ncei.noaa.gov/access/monitoring/scec/records/all/tmax

A NOAA graph entitled "Contiguous U.S. Maximum Temperature" shows the U.S. maximum temperatures between 1985 and 2024. It's clear from this graph that there is no upward trend in maximum temperatures.

If the planet was indeed boiling, we would probably see an upward trend in the past 120+ years. However, what we do see is less cold in the winters or milder winters, which is happening throughout the Northern Hemisphere, including the Arctic according to the two graphs below.

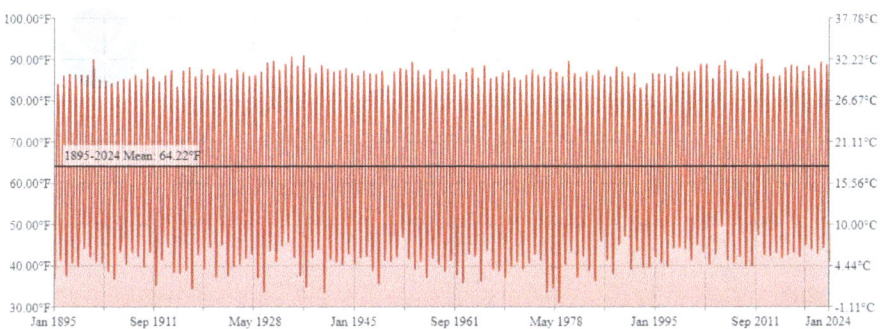

It is wise to note that in the second graph, also provided by NOAA, even the milder winters do not change significantly; but according to winter minimums at the bottom of the graph, you can see them grow slightly warmer, almost imperceptibly[71].

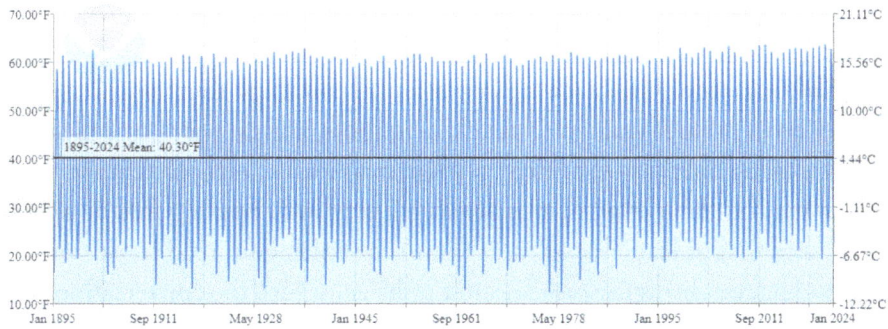

---

71   https://www.ncei.noaa.gov/access/monitoring/climate-at-a-glance/national/time-series/110/tmin/1/0/1895-2024?base_prd=true&begbaseyear=1895&endbaseyear=2024

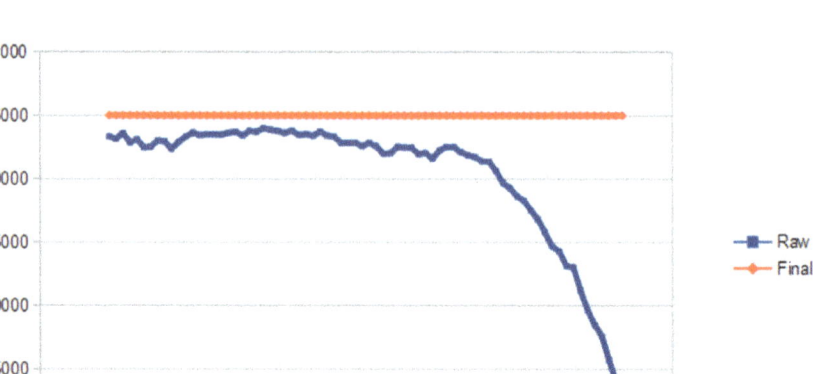

The graph above is made by counting the number of reported monthly temperatures in the final and raw data sets. NOAA has lost 30% of their station data since 1990, but still reports adjusted temperatures for the missing data.

https://realclimatescience.com/2014/06/more-than-40-of-ushcn-station-data-is-fabricated/#gsc.tab=0

Although the climate alarmists warn of record high temperatures, these simply don't exist. If you have friends, neighbors, family, peers, or colleagues who are being bullied and beaten up about heat waves and high temperatures, please hand them a copy of this book. If everybody reading this book will make sure that at least five of their friends or colleagues read it too, we will have a chance at bringing more honest politicians into the conversation who aren't making decisions for their corporate interests, as opposed to the good of their constituents and the Earth itself.

THE POLAR VORTEX

The polar vortex is used as a propaganda tool, in the same way heat waves were leveraged as described in the previous pages.

During the 1970s, there were predominant wavy polar vortex patterns which were blamed on global cooling. Hilariously, those same wavy polar vortex patterns are now blamed on global warming! Same data, different spin.

In the image below from Climate.gov, we read in a graphic entitled "Understanding the Polar Vortex" we read:

> *"The Arctic polar vortex is a strong band of winds in the stratosphere, surrounding the North Pole 10-30 miles above the surface. The polar vortex is far above and typically does not interact with the polar jet stream, the flow of winds in the troposphere 5-9 miles above the surface. But when the polar vortex is especially strong and stable, the jet stream stays farther north and has fewer 'kinks.' This keeps cold air contained over the Arctic and the mid-latitudes warmer than usual. Every other year or so, the Arctic polar vortex dramatically weakens. The cortex can be pushed off the pole or split into two. Sometimes the polar jet stream mirrors this stratospheric upheaval, becoming weaker or wavy. At the surface, cold air is pushed southward to the mid-latitudes, and warm air is drawn up into the Arctic[72]."*

---

[72] https://www.climate.gov/news-features/understanding-climate/understanding-arctic-polar-vortex

Understanding the polar vortex

The Arctic polar vortex is a strong band of winds in the stratosphere, surrounding the North Pole 10–30 miles above the surface.

The polar vortex is far above and typically does not interact with the polar jet stream, the flow of winds in the troposphere 5–9 miles above the surface. But when the polar vortex is especially strong and stable, the jet stream stays farther north and has fewer "kinks." This keeps cold air contained over the Arctic and the mid-latitudes warmer than usual.

Every other year or so, the Arctic polar vortex dramatically weakens. The vortex can be pushed off the pole or split into two. Sometimes the polar jet stream mirrors this stratospheric upheaval, becoming weaker or wavy. At the surface, cold air is pushed southward to the mid-latitudes, and warm air is drawn up into the Arctic.

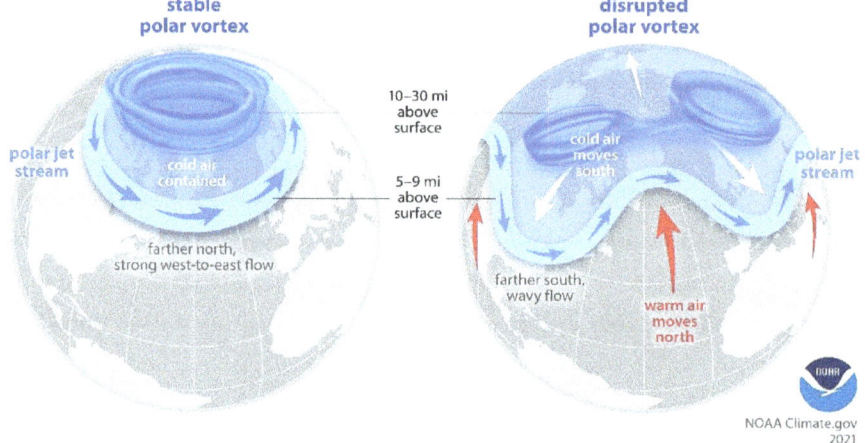

Photo Courtesy of Climate.gov

In truth, we don't understand why we have a stable polar vortex or a wavy polar vortex. We do know that they vary just like the La Niña and El Niño Pacific Ocean weather patterns. This is important because when there is a wavy polar vortex, there are what appear to be fingers of cold air moving south and fingers of very warm air moving north.

The propaganda media will *only* tell us about the hot temperatures in the north, even though right next door there are very cold temperatures that are not reported.

Part Four:
DEBUNKING OCEAN LIES

CHAPTER THIRTEEN:
CHANGING SEA LEVELS

> *"The IPCC notes that no significant acceleration in the rate of sea level rise during the 20th century has been detected. This did not appear in the IPCC Summary for Policymakers."*
> Robert C. Balling, Jr.

Robert C. Balling, Jr. is a professor of geography at Arizona State University and the former director of its Office of Climatology. His research primarily focuses on climatology, global climate change, and geographic information systems. Balling is recognized for his skepticism concerning man-made global warming claims.

In his 2009 book *Heated Debate: Greenhouse Predictions Versus Climate Reality*, Balling states:

> *"Sea levels have been rising unevenly across the world since the 1850s, coinciding with the end of the Little Ice Age, also known as pre-industrial times. Reliable tidal records from around the world date back to about 1880, providing valuable insights into this phenomenon.*
>
> *One key factor contributing to rising sea levels is the land rising out of the ocean. Contrary to popular belief, not all coastal areas are experiencing sea level rise. Many regions witness land uplift, caused by Earth plate shifts and movement, causing minimal change in ocean levels. It's essential to understand that ocean levels are measured relative to the land where the measurement occurs.*

> *In the following graph you'll see that the global sea level rise has been increasing at about eight inches per century, which is a little less than an inch per decade. There is no evidence that this is accelerating.*
>
> *In some regions like Indonesia, excessive groundwater extraction has led to land subsidence, exacerbating the effect of rising sea levels. Similarly, the East Coast of the United States experiences more land sinking compared to the West Coast or Alaska."*

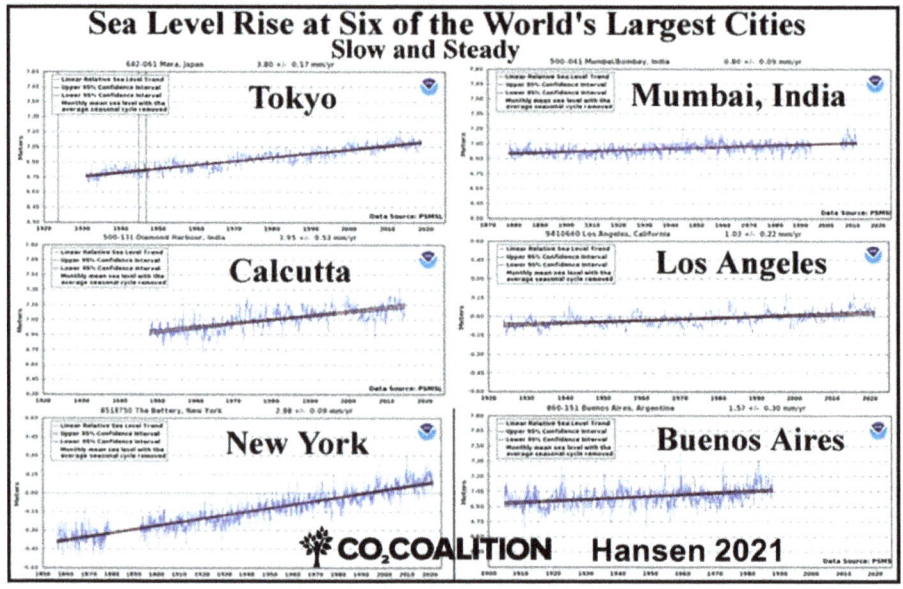

The Associated Press article below from June 30, 1989 states, "A senior U.N. environmental official says entire nations could be wiped off the face of the Earth by rising sea levels if the global warming trend is not reversed by the year 2000."

U.N. Predicts Disaster if Global Warming Not Checked

PETER JAMES SPIELMANN   June 30, 1989

UNITED NATIONS (AP) _ A senior U.N. environmental official says entire nations could be wiped off the face of the Earth by rising sea levels if the global warming trend is not reversed by the year 2000.

Photo Courtesy of The Associated Press

The same article states, "Coastal flooding and crop failures would create an exodus of 'eco- refugees,' ...threatening political chaos," according to Noel Brown, director of the New York office of the U.N. Environment Program, or UNEP.

Brown says in the article that governments had a 10-year window of opportunity to solve the greenhouse effect before it got out of human control. The article continues to state that as global warming melts polar ice caps, ocean levels would rise by up to three feet, enough to cover the Maldives and other flat island nations, according to Brown from an interview by the AP.

> *"Coastal regions will be inundated; one-sixth of Bangladesh could be flooded, displacing a fourth of its 90 million people. A fifth of Egypt's arable land in the Nile Delta would be flooded, cutting off its food supply, according to a joint UNEP and U.S. Environmental Protection Agency study... Ecological refugees will become a major concern, and what's worse is you may find that people can move to drier ground, but the soil and the natural resources may not support life. Africa doesn't have to worry about land, but would you want to live in the Sahara?"*

Bangladesh's population has surged to 174 million people, and the incidence of poverty has significantly decreased. Additionally, Bangladesh has experienced an expansion in landmass, with approximately 230 square miles or nearly 600 square kilometers of new land formed. Remarkably, this increase in land area surpasses the total landmass of the Maldives[73].

Now, thirty-five years later we know that all of these catastrophic predictions have proven inaccurate.

In numerous oceanic islands, a process called accretion occurs, wherein sand and coral growth contribute to the expansion of the landmass. Conversely, countries like Sweden have seen sea levels drop due to land uplift caused by tectonic plate movement. Nearby countries like Norway, however, experience stable sea levels. In the *TIME Magazine* cover below, you'll see António Guterres, Former Socialist Prime Minister of Portugal, and U.N. Secretary General since 2017, standing in water with the headline, "Our Sinking Planet" on a cover from 2019.

This magazine cover is an excellent example of propaganda; the UN. Secretary General donning a suit standing thigh deep in the ocean, of course looking very dismal. The caption over his chest says "Rising Seas. Fleeing Residents. Disappearing Villages."

https://apnews.com/article/bd45c372caf118ec99964ea547880cd0

In the article, Guterres falsely claims that the world is now boiling and advocates granting the U.N. control over nations, claiming that he knows how to save the planet. Conveniently, his plan involves forcing global citizens to surrender their money, not to mention their freedoms.

As mentioned, I've spent time in the belly of the beast as a politician. I am certain and, frankly, terrified, that if we don't prevent the U.N.

---

73 https://servir.icimod.org/news/natural-coastal-land-expansion-offers-hope-to-low-lying-bangladesh/

from gaining even more authority over our nation and others, we're on a path to global slavery.

U.S. Congressman John E. Rankin said, "The United Nations is the greatest fraud in history. Its purpose is to destroy the United States." There is a myriad of reasons why Rankin and others pinpoint the U.N. among the most significant enemies of a prosperous future for global citizens.

In fact, the U.N. World Constitution states, "The age of nations must end. The governments of nations have decided to order their separate sovereignties into one government to which they will surrender their arms."

The U.N. is riddled with corruption; it's an unelected body that's primary focus is to usher in a "one world" government and "new world order." Since the U.N.'s inception, there has been enough corruption to merit its own book. Among other crimes, U.N. efforts have led to the disasters in Sri Lanka and South Africa.

Back to the rising sea levels issue, contrary to alarmists' claims, cities like New York remain unscathed by rising sea levels. Despite predictions made by figures like James Hansen in 1989, the city remains unaffected. The tidal record for New York City at the station closest shows the sea level trend observed over the past century. In the image below, you'll see that sea levels around the Statue of Liberty in the past 100 years have remained steady, neither rising nor falling significantly.

Interestingly, many affluent people who are most adamantly promoting the climate change agenda also own beachfront properties, like former President Barack Obama. This implies that the Obamas are not genuinely concerned about rising sea levels; at least not in the near future.

Historical predictions of catastrophic sea level rise, such as those made by the U.N. Environment Program, have consistently failed to materialize.

Countries like Bangladesh, once identified as being at risk of submersion, have experienced significant landmass growth.

Recent reports, such as one from NASA in 2022, suggest significant sea level rise in the coming decades, "As much as 12 inches (30 cm) by 2050[74]." However, these projections exaggerate the steady 8/10s of an inch (2 cm) per decade that has been the constant. NOAA's prediction lacks scientific basis. The reality is that sea level rise has been gradual and manageable, with minimal impact on coastal regions.

In this graph of Stockholm Sweden tides and currents, NOAA estimates that the ocean level is actually going down at a rate of about 1.2 feet a century.

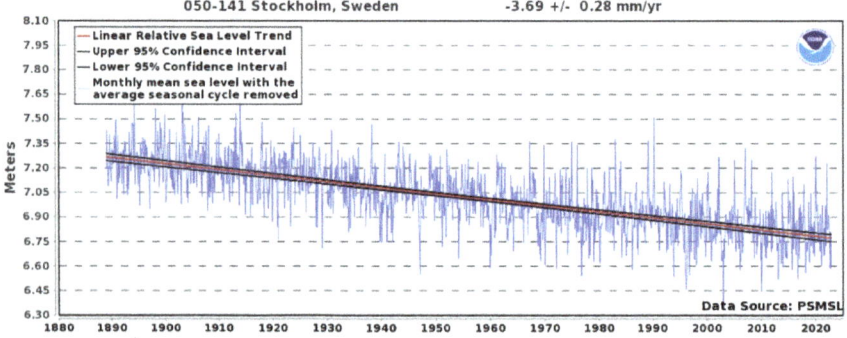

Source https://tidesandcurrents.noaa.gov/sltrends/sltrends_station.shtml?id=050-141

In Sydney, Australia, they estimate the trend to be a quarter foot every 100 years in the following NOAA tides and currents graph.

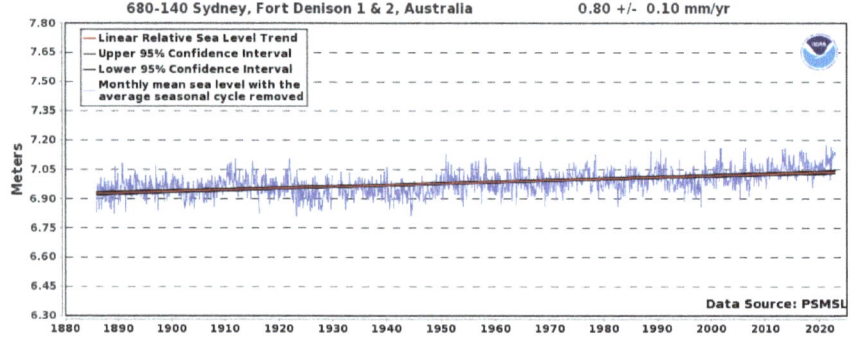

Source https://tidesandcurrents.noaa.gov/sltrends/sltrends_station.shtml?id=680-140

[74] https://www.forbes.com/sites/priyashukla/2022/02/25/sea-levels-could-rise-12-inches-by-2050-says-new-federal-report/?sh=5b1a92d308f4

In Antofagasta, Chile, sea level is declining by about a quarter foot per century, as shown in the following NOAA graph.

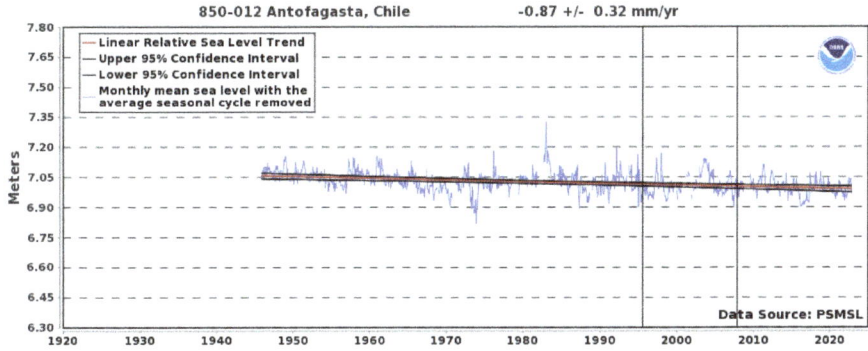

Source https://tidesandcurrents.noaa.gov/sltrends/sltrends_station.shtml?id=850-012

Studies debunking alarmist claims highlight the phenomenon of "landification" or "land emergence," wherein coastal land areas are expanding rather than receding. Despite mainstream media silence on this topic, research indicates that accretion outweighs erosion in coastal areas globally, meaning that even with sea level rise, the total coastal land area is increasing.

"Even with rising sea levels, the world's coastal land area has increased by 13,000 square miles or about the size of Belgium!" says Dr. Roger Pielke Jr. in his Substack article titled, "Landification,"[75] where he identifies many coastal areas around the world that are actually growing.

Below, you'll see that at almost every latitude and longitude, landification is winning out over erosion and sea level rise or "SLR[76]." The following graph shows areas losing ground to the ocean and sinking in red, and the green areas that are growing and adding land are more plentiful. Some areas are sinking like the East Coast of the U.S. Other areas are rising, like the Pacific Northwest and Sweden.

---

[75] https://rogerpielkejr.substack.com/p/landification
[76] https://www.sciencedirect.com/science/article/abs/pii/S0924271621002598?via%3Dihub

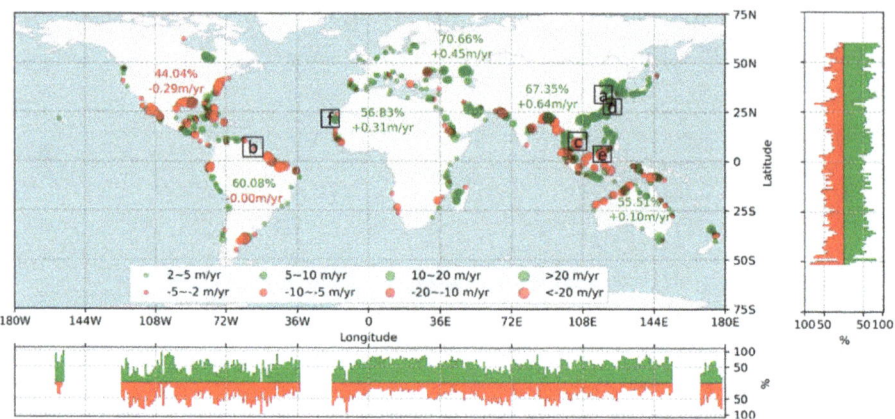

Image Courtesy of ScienceDirect.com

You can check out the estimated sea level rise for a variety of U.S. locations at the website https://tidesandcurrents.noaa.gov/sltrends/msltrendstable.htm. Sea level has decreased the most in Skagway, Alaska by 6.7 inches (171 cm), second most in Juneau, Alaska by 5 inches (129 cm), and increased the most at Eugene Island, Louisiana by 3.8 inches (96.5 cm).

WHERE IS THE WATER?

You might be asking the question, where is Earth's water? That is the title of the image below courtesy of the U.S. Geological Survey Water Science School.[77] In this image, we learn that the vast majority of Earth's water is found in the oceans, accounting for 96.5% of the total water. About two-thirds of the remaining 2.5% freshwater is stored in glaciers and ice caps (about 1.65% of all the water on Earth is glacial ice and polar ice caps). Because the North Pole sea ice is already in the ocean, it would not raise sea levels if it melted.

You can test this yourself. Take a glass of water with some ice cubes in it. Measure the water level. Let the ice melt, and you will find that the water level has not increased. While frozen water is larger (when expanded) than

---

[77] https://www.usgs.gov/special-topics/water-science-school/science/where-earths-water

liquid water, it does not increase the water level when it melts again. Because ice is lighter, it floats.

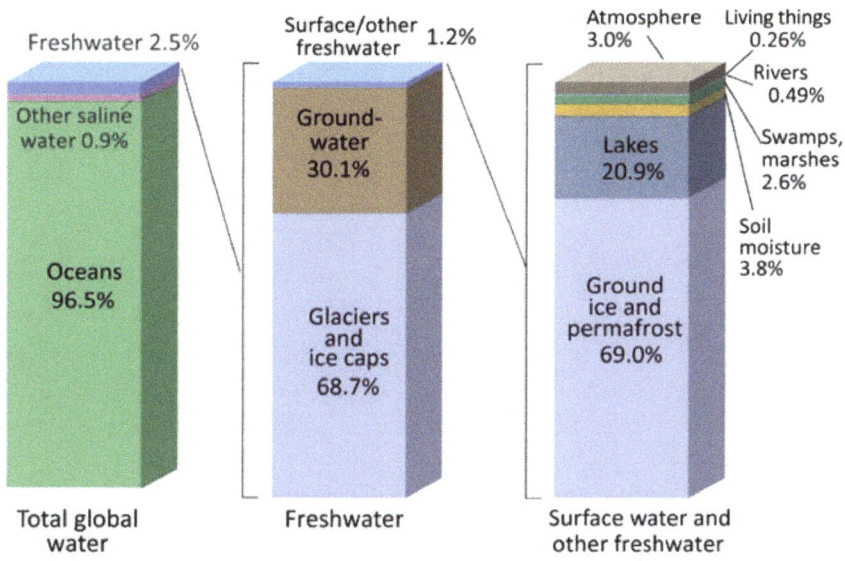

Photo Courtesy of the United States Geological Survey

The Greenland glacial sheet, which is miles thick, would take thousands of years to completely melt due to its immense size. Such an event is highly unlikely because of the perpetually freezing temperatures in Greenland.

Similarly, Antarctica, where over 89% of Earth's ice is located, has seen no warming in over seventy years. In fact, Antarctica has even set new cold records in recent years. There isn't the warming claimed in either of the poles to justify the claims of exponential sea level rise. It doesn't make sense that current trends should dramatically change, when there are not dramatic temperature shifts. Antarctica, which has the vast majority of the world's ice, is just as cold as it was 70 years ago.

CHAPTER FOURTEEN:
CORAL REEFS AND ISLAND EXPANSION

"The Great Barrier Reef is doing fine..."
Dr. Walter Starck

Dr. Walter Starck is one of the pioneers in the scientific investigation of coral reefs. He grew up in the Florida Keys and received a Ph.D in Marine Science from the University of Miami in 1964. He has over 40 years' experience in worldwide reef studies and his work has encompassed the discovery of much of the basic nature of reef biology.

Starck is behind the discovery of over one hundred species of fish which were previously unknown to scientists as well as the discovery of numerous corals, shells, crustaceans, and other new discoveries. Starck is a reef expert, and he has no fear that climate change is destroying reefs. I tend to take his word a bit more seriously than your college professor or colleague who pontificates about reefs they've only experienced while snorkeling a handful of times on vacation, coupled with propaganda about the reef decimation from a biased source who state that, "the current 'dire threats' to the Reef involve coral bleaching, attributed to climate change and declining water quality said to be a consequence of farming and grazing[78]." According to Starck: "The real-world evidence presents a quite different picture[79]."

In another article he states that, "ongoing research elsewhere is finding that evidence of bleaching events is common in coral skeletons going back

---

[78] https://www.goldendolphin.com/wstarck.htm
[79] https://www.goldendolphin.com/wstarck.htm

hundreds of years and rapid recovery is also the normal outcome. The frequency and severity of such events varies in different localities with no statistically significant trend of increase or decrease over time[80]." This quote is written in the article "Simple Clear Evidence the Great Barrier Reef is Doing Fine," "where evidence is presented that shows that so-called 'reef alarmists' are beating drums based upon unfounded hypotheticals about the reefs; they're, in essence, looking to achieve their own fifteen minutes of fame in a world of climate alarmism."

From NOAA's Ocean Education service: "Coral continue to require warm water and thrive in the warmest of Earth's waters... The majority of reef building corals are found within tropical and subtropical waters. These typically occur between 30° north and 30° south latitudes. The red dots on this map show the location of major stony coral reefs of the world[81]."

Coral reefs are ancient ecosystems that have endured for millions of years and continue to thrive despite fluctuations in Earth's climate. While the narrative surrounding their health includes dire predictions, let us look deeper at the truth.

CORAL REEFS ARE DOING JUST FINE

Coral reefs predominantly inhabit warm waters, with tropical and subtropical regions hosting the majority of these vibrant ecosystems. This preference for warm waters is reflected in their distribution across the globe, as highlighted by NOAA's mapping of major stony coral reefs in the graph below where the reefs are concentrated mostly around the equator and Pacific Ocean, with many also found off of Asia and Australia.

Although some deep-sea corals do exist in colder waters, they're a minority and are not a primary focus of conservation efforts.

---

[80] https://quadrant.org.au/opinion/doomed-planet/2021/05/simple-clear-evidence-the-great-barrier-reef-is-fine/

[81] https://oceanservice.noaa.gov/education/kits/corals/media/supp_coral05a.html

Chapter Fourteen: Coral Reefs and Island Expansion

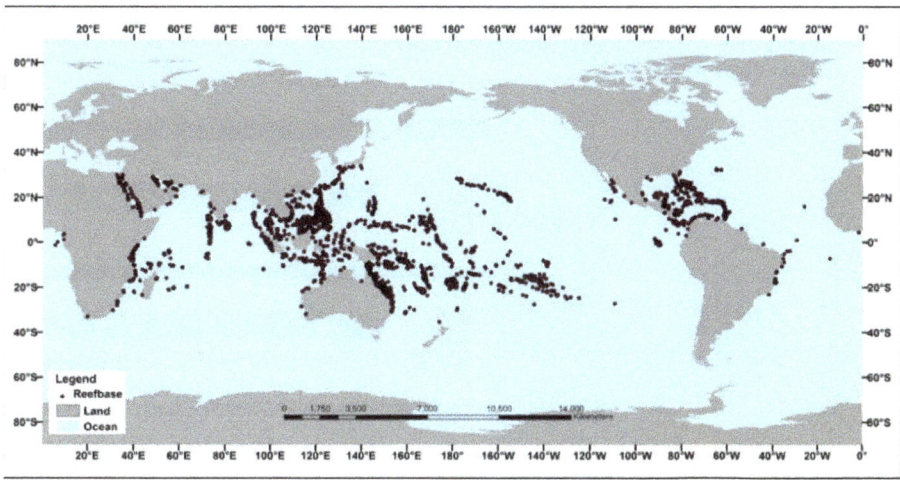

NOAA[82]

Central to the success of coral reefs is the symbiotic relationship between coral polyps and Zooxanthellae algae. These organisms rely on each other for survival; with the algae providing food and vibrant colors to the coral while the coral provides shelter and nutrients in return.

Coral bleaching, often attributed to global warming, is actually completely healthy and natural. It's a phenomenon that has occurred throughout history when Zooxanthellae abandon coral polyps. Despite the alarmist narrative surrounding coral bleaching, these events are part of the reef's natural cycle, with corals typically recovering as new algae colonize the polyps.

Contrary to popular belief, coral reefs, including the iconic Great Barrier Reef, are thriving. The doom-laden predictions of climate alarmists are false and a complete denial of reality. In fact, as you'll see in the graph below from the Australian Institute of Marine Science, the coral of the Great Barrier Reef is at an all-time high[83].

---

[82] https://oceanservice.noaa.gov/education/tutorial_corals/media/supp_coral05a.html
[83] https://www.aims.gov.au/reef-monitoring/gbr-condition-summary-2020-2021

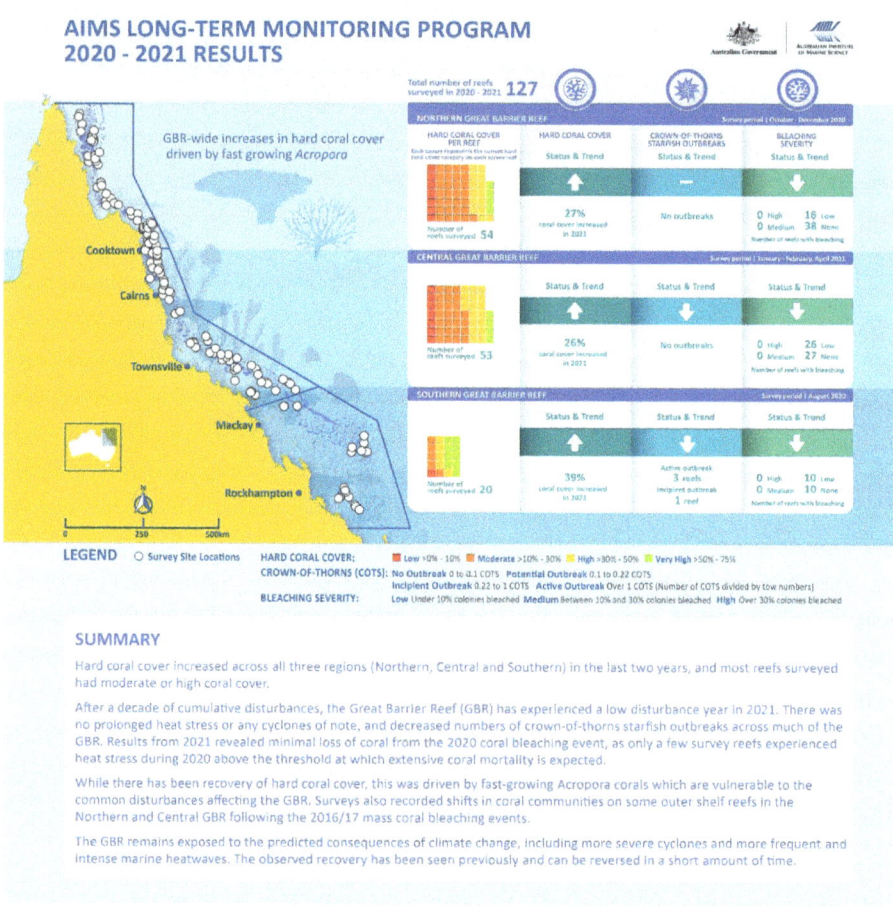

Photo Courtesy of The Australian Institute of Marine Science Long-Term Monitoring Program

Despite this empirical evidence, the U.N. still attempts to designate coral reefs as endangered. This makes Australia and other countries who prioritize internal versus external control of reef management uneasy. After all, who wants a group of overweight delegates making decisions about reefs when the closest they've ever been to one is from a booze cruise or private yacht?

These nations are right to feel hesitant about the U.N. asserting control through the guise of climate change in order to expand their power. Who voted for the U.N., anyway? Oh, that's right. Nobody. It's an entitled and unelected group of global elites vying for power and control over nations and their sovereignty and, in this case, co-opting their well-earned wisdom.

Chapter Fourteen: Coral Reefs and Island Expansion

While it is true that coral reefs face challenges such as pollution, overfishing, and mass tourism, climate change is not a threat to reefs. The equatorial regions of Earth are barely warming because of climate change. The portrayal of reefs as victims of climate change neglects their ability to adapt and thrive in warm waters, where they flourish with adequate nutrients and protection.

Here are examples of the coral reef propaganda that they routinely blame on climate change from a 2024 DuckDuckGo search:

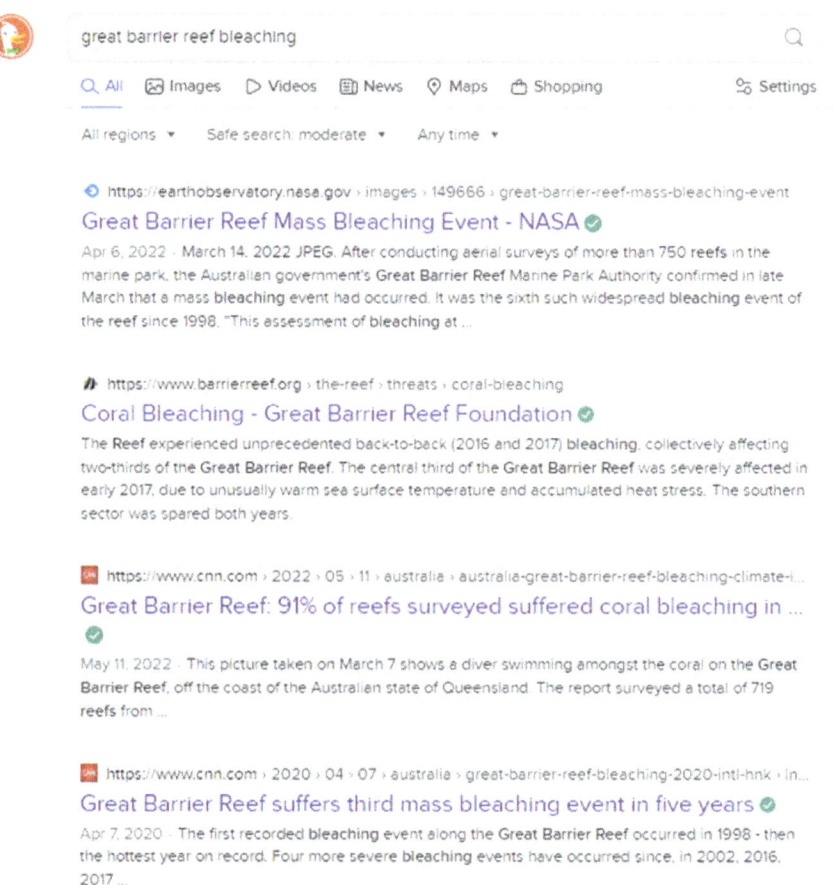

In conclusion, the propaganda of warmer water being the cause of coral reef decline due to man-made climate change overlooks the resilience and adaptability of these ecosystems. It ignores the fact that coral thrives in warmer water, providing the right nutrients and circulation.

By prioritizing responsible management and protection from real threats like pollution and habitat degradation, we can ensure the continued success of coral reefs for generations to come. Except for deep sea varieties, coral around the world prefers warmer waters, including the Great Barrier Reef.

THE TRUTH ABOUT OCEAN ISLANDS

> *"A 2019 global-scale analysis of 709 islands in the Pacific and Indian Oceans revealed 89% were either stable or growing in size, and that no island larger than 10 hectares had decreased in size since the 1980s."*
> Virginie Duvat-Magnan

Virginie Duvat-Magnan is a Professor of Geography and a researcher in the Littoral Environenment et Sociétés (LIENSs) laboratory[84]. Her expertise lies in geomorphology, focusing on the study of the processes shaping islands, with a particular emphasis on the risks associated with coastal areas and development.

Her research extends to examining the impacts of climate change on tropical coastlines and the management of coastal environments. Duvat-Magnan is internationally recognized for her specialization in tropical islands. She wrote the "Small Island" chapter of the 5th report of the Intergovernmental Panel on Climate Change (IPCC), published in 2014.

Since 2012, Duvat-Magnan has been a member of the scientific committee of the National Strategy for Integrated Coastline Management at the Ministry of the Environment. She has also been appointed by the Institute of Ecology and the Environment of the CNRS to the Multidisciplinary Thematic Network Littoral.

Her research findings have been disseminated through nearly 200 scientific publications and communications, including over fifty scientific articles in national and international journals, approximately forty books

---

[84] https://lienss.univ-larochelle.fr/

Chapter Fourteen: Coral Reefs and Island Expansion

and book chapters, and more than 100 conference presentations at both national and international levels

In the image below, you'll see three data points about island sinking provided by ResearchGate.net[85], Notrickszone.com[86], and PSMSL.org[87]. The data points are in response to a *Washington Post* article entitled, "A sinking nation is offered an escape route. But there's a catch." The article details the stories of several inhabitants of the nation of Tuvalu who've been convinced their island home in the Pacific is sinking. Of course, the U.N. is heavily involved in Tuvalu propaganda and is mentioned in this article[88].

The main three data points that counter the terrifying article about the nation of Tuvalu sinking include the following:

1. Tuvalu is not sinking: "Places like Tuvalu, Kiribati and Vanuatu - all notorious for an inferred sea level rise - have tide gauges, which show no ongoing sea level rise[89]."
2. Tuvalu is actually *gaining* land area.[90]
3. Tidal gauge data at Funafuti doesn't show any special rise in sea level[91].

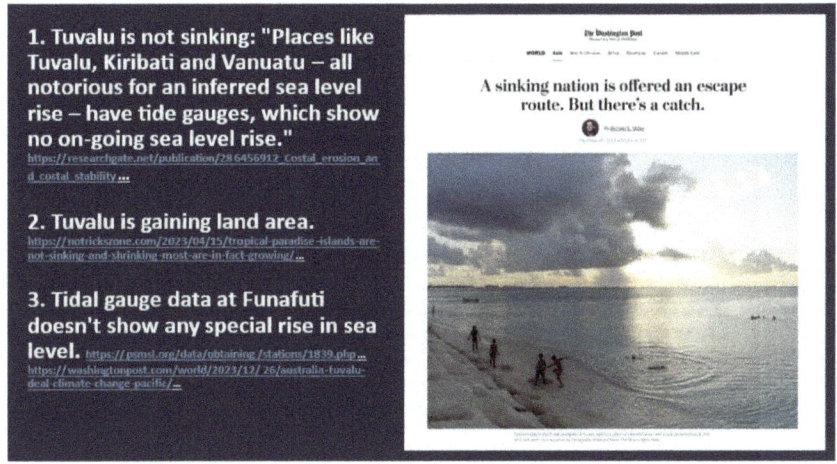

---

[85] https://www.researchgate.net/publication/286456912_Costal_erosion_and_costal_stability
[86] https://notrickszone.com/2023/04/15/tropical-paradise-islands-are-not-sinking-and-shrinking-most-are-in-fact-growing/
[87] https://psmsl.org/data/obtaining/stations/1839.php
[88] https://www.washingtonpost.com/world/2023/12/26/australia-tuvalu-deal-climate-change-pacific/
[89] https://www.researchgate.net/publication/286456912_Costal_erosion_and_costal_stability
[90] https://wattsupwiththat.com/2023/04/15/tropical-paradise-islands-are-not-sinking-and-shrinkingmost-are-in-fact-growing/
[91] https://psmsl.org/data/obtaining/complete.php

Contrary to the narrative portrayed by the media and political agendas, many ocean islands just like Tuvalu are *not* sinking beneath rising seas. In fact, numerous scientific studies reveal that 89% of these islands are either stable or expanding in size. While headlines often sensationalize the plight of sinking islands and fleeing residents, the reality is that most islands are growing, while only a few are shrinking.

As we've detailed in this book, for over three decades now (recall the 1988 article that detailed James Hansen's hysteria), we've been repeatedly warned by the U.N. and scientists who've likely been paid, that "entire nations could be wiped off the face of the Earth by rising sea levels if the global warming trend is not reversed[92]."

Studies conducted over several decades using aerial photographs dating back to the mid-20th century have consistently shown that the majority of reef islands in the equatorial Pacific, Micronesia, and the Gilbert Islands are experiencing net shoreline expansion. Similarly, research focused on the Atoll Islands in the tropical Pacific and Indian oceans has revealed predominantly stable or accretionary trends in island area. The image below illustrates growth on Tuvalu from 1971 to 2014, which is consistent with 89% of ocean islands[93].

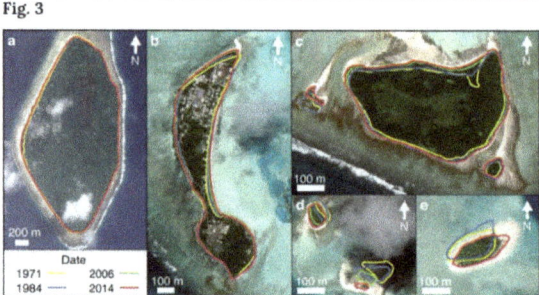

Fig. 3

Examples of island change and dynamics in Tuvalu from 1971 to 2014. **a** Nanumaga reef platform island (301 ha) increased in area 4.7 ha (1.6%) and remained stable on its reef platform. **b** Fangaia island (22.4 ha), Nukulaelae atoll, increased in area 3.1 ha (13.7%) and remained stable on reef rim. **c** Fenualango island (14.1 ha), Nukulaelae atoll rim, increased in area 2.3 ha (16%). Note smaller island on left Teafuafatu (0.29 ha), which reduced in area 0.15 ha (49%) and had significant lagoonward movement. **d** Two smaller reef islands on Nukulaelae reef rim. Tapuaelani island, (0.19 ha) top left, increased in area 0.21 ha (113%) and migrated lagoonward. Kalilaia island, (0.52 ha) bottom right, reduced in area 0.45 ha (85%) migrating substantially lagoonward. **e** Teafuone island (1.37 ha) Nukufetau atoll, increased in area 0.04 ha (3%). Note lateral migration of island along reef platform. Yellow lines represent the 1971 shoreline, blue lines represent the 1984 shoreline, green lines represent the 2006 shoreline and red lines represent the 2014 shoreline. Images ©2017 DigitalGlobe Inc

---

[92] https://apnews.com/article/bd45c372caf118ec99964ea547880cd0
[93] https://notrickszone.com/2019/01/17/new-science-89-of-the-globes-islands-and-100-of-large-islands-have-stable-or-growing-coasts/

EVIDENCE OF ISLAND EXPANSION

The Maldives, a proverbial poster child for the purported threat of rising sea levels, has actually seen an expansion of its islands by 37.5 square kilometers over a period of just seventeen years. Furthermore, a global-scale analysis of hundreds of islands in the Pacific Ocean and Indian Ocean found that nearly 90% were either stable or growing in size since the 1980's[94].

Teams of scientists published a series of studies including aerial photographs from the 1940s, 1960s, and 1970s which show a net shoreline expansion in the equatorial Pacific, Micronesia, and Gilbert Islands[95,96].

A 2019 global-scale analysis of 709 islands in the Pacific and Indian oceans shows nearly 90% were either stable or growing, and that no "larger" island has shrunk since the 1980s[97].

And yet another study illustrates trends post 2000, that shorelines of hundreds of Pacific and Indian Ocean islands are expanding, and they continue to grow[98].

These scientific findings debunk the notion that entire nations are on the brink of extinction due to climate change-induced sea level rise. Instead, they highlight the resilience and dynamic nature of ocean islands in the face of changing environmental conditions. That's something to celebrate!

It is important to remember: we're being propagandized for a purpose; and the Earth is not the beneficiary in this web of climate change gaslighting, the beneficiaries are the shareholders from the U.N., WHO, and WEF.

---

[94] https://www.sciencedirect.com/science/article/abs/pii/S2213305421000059
[95] https://www.sciencedirect.com/science/article/abs/pii/S0169555X20305572
[96] https://www.sciencedirect.com/science/article/abs/pii/S0169555X21002397
[97] https://wires.onlinelibrary.wiley.com/doi/abs/10.1002/wcc.557
[98] https://www.sciencedirect.com/science/article/abs/pii/S2213305421000059

THE TRUTH ABOUT OCEAN PH AND ACIDIFICATION

Ian Rutherford Plimer is a prominent Australian geologist and academic. He has been a professor of mining geology at the University of Adelaide and later as a professor emeritus at the University of Melbourne. He has made significant contributions to the fields of mineral exploration and mining.

In one of his books, *Not for Greens: He Who Sups with the Devil Should Have a Long Spoon*, Plimer explores environmentalism's societal impact. Another book titled, *A Short History of the World* offers an overview of geological and historical events. Plimer actively participates in public discourse on geology, climate change, and environmental policy. He's a vocal critic of mainstream climate science, advocating for a more skeptical approach to environmental issues.

Overall, Plimer is widely respected in geology, having made significant contributions through academia, authorship, and public engagement. After being questioned during his time at the University of Adelaide, Pilmer states:

> "The oceans have remained alkaline during the Phanerozoic (last 540 million years) except for a very brief and poorly understood time 55 million years ago... Rainwater (pH 5.6) reacts with the most common minerals on Earth (feldspars) to produce clays, this is an acid consuming reaction, alkali and alkaline earths are leached into the oceans (which is why we have saline oceans), silica is redeposited as cements in sediments, the reaction consumes acid and is accelerated by temperature (see below). In the oceans, there is a buffering reaction between the seafloor basalts and sea water (see below). Sea water has a local and regional variation in pH (pH 7.8 to 8.3). It should be noted that pH is a log scale and that if we are to create acid oceans, then there is not enough CO_2 in fossil fuels to create oceanic acidity because most of the planet's CO_2 is locked up in rocks. When we run out of rocks on Earth or plate tectonics ceases, then we will have acid oceans[99]."

---

[99] https://jennifermarohasy.com/2008/10/not-enough-CO2-to-make-oceans-acidic-a-note-from-professor-plimer/

Contrary to what climate alarmists and media outlets claim, the Earth's oceans are not becoming more acidic due to increasing carbon dioxide levels. In fact, the oceans are naturally base, meaning they have a pH above 7

The reason oceans are base is because of dissolved minerals and salts within them. Their pH typically ranges from about 7.8 to 8.5, with an average around 8.1. This is far from being acidic, as neutral pH is 7 and anything above that is considered base.

See the figure below to compare the pH of common substances[100].

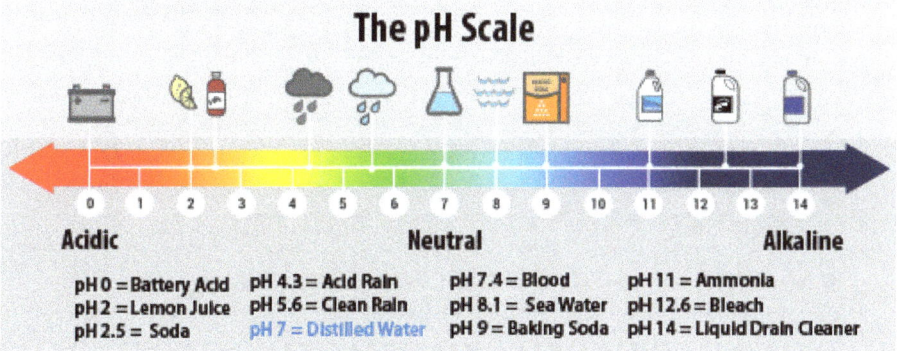

Photo Courtesy of the U.S. Environmental Protection Agency, "Measuring Acid Rain," epa.gov, last accessed February 22, 2023

While some organizations like NASA and NOAA propagate the narrative of ocean acidification, it's clear that such claims are based on hypothetical scenarios and models, and they lack actual measured data.

---

[100] https://www.epa.gov/acidrain/what-acid-rain

> https://oceanservice.noaa.gov › facts › acidification.html
> **What is Ocean Acidification? - National Ocean Service** ✓
> When CO 2 is absorbed by seawater, a series of chemical reactions occur resulting in the increased concentration of hydrogen ions. This increase causes the seawater to become more acidic and causes carbonate ions to be relatively less abundant. Carbonate ions are an important building block of structures such as sea shells and coral skeletons.
>
> https://climatekids.nasa.gov › acid-ocean
> **What Is Ocean Acidification? | NASA Climate Kids** ✓
> Dec 20, 2022 · Because ocean water has become more acidic, some animals — like certain oysters and clams — are having difficulty in making or keeping their shells. For example, acidic ocean water can cause coral to grow more slowly and weaken coral reefs. These reefs are an important home for many living things. Their health is essential to many ecosystems.
>
> https://www.nationalgeographic.com › environment › article › critical-issues-ocean-acidification
> **Ocean acidification facts and information - Environment** ✓
> The oceans are growing more acidic, and scientists think the change is happening faster than at any time in geologic history. That's bad news for most creatures that live in the ocean, many...
>
> https://ocean.si.edu › ocean-life › invertebrates › ocean-acidification
> **Ocean Acidification | Smithsonian Ocean** ✓
> The pH scale goes from extremely basic at 14 (lye has a pH of 13) to extremely acidic at 1 (lemon juice has a pH of 2), with a pH of 7 being neutral (neither acidic or basic). The ocean itself is not actually acidic in the sense of having a pH less than 7, and it won't become acidic even with all the CO 2 that is dissolving into the ocean.
>
> https://www.britannica.com › science › ocean-acidification
> **Ocean acidification | Definition, Causes, Effects, Chemistry, & Facts** ✓
> ocean acidification, the worldwide reduction in the pH of seawater as a consequence of the absorption of large amounts of carbon dioxide (CO2) by the oceans. Ocean acidification is largely the result of

Photo Courtesy of DuckDuckGo search from November 18th 2023 and verified with similar results on March 12, 2024. Google adds an NRDC description, which is a climate alarmist organization.

Studies have shown that increased carbon dioxide levels in the ocean actually benefit marine life, particularly phytoplankton, which is the foundation of the ocean food chain. More CO_2 leads to enhanced phytoplankton growth, contributing to a healthier ocean ecosystem, just the same as it does on land, which we learned about in Part Two of this book.

Because oceans don't actually become more acidic, it's more accurate to describe any potential changes in ocean pH as a reduction in baseness rather than an increase in acidity. While rainwater, with a pH of about 5.6, is acidic, the oceans remain firmly in the base category.

The notion of ocean acidification is not truthful; it's being used to promote a narrative, but has no base in reality. A simple understanding of the natural base conditions of the oceans and the potential benefits of increased carbon dioxide levels should put to rest concern about increasingly acidic oceans.

ADDITIONAL RESOURCES

1. Pacific Marine Environmental Laboratory, "What is Ocean Acidification?," National Oceanic and Atmospheric Administration, accessed August 12, 2021, https://www.pmel.noaa.gov/CO2/story/What+is+Ocean+Acidification%3F

2. K. Caldeira and M.E. Wickett, "Ocean Model Predictions of Chemistry Changes from Carbon Dioxide Emissions to the Atmosphere and Ocean," Journal of Geophysical Research, Volume 110, September 21, 2005, https://agupubs.onlinelibrary.wiley.com/doi/10.1029/2004JC002671

3. CO_2 Coalition, Ocean Health: Is There an 'Acidification' Problem?, June 2020, accessed August 12, 2021, http://CO2coalition.org/wp-content/uploads/2020/06/Steele-Ocean- Health-White-Paper-final-5-28-20.pdf

CHAPTER FIFTEEN:
GLACIERS & NATURAL CYCLES

"This number (of receding glaciers reported by the IPCC) is not just a little bit wrong, but far out of any order of magnitude ... It is so wrong that it is not even worth discussing."
Georg Kaser

Georg Kaser is a prominent figure in climate research, renowned for his contributions to glaciology and his influential role as a lead author for the Intergovernmental Panel on Climate Change (IPCC). With a background in meteorology and geography from the Leopold Franzens University of Innsbruck, Kaser's expertise spans diverse topics within the field of climate science.

His doctoral work focused on the evaporation processes of snow and ice, followed by research on tropical glaciers. As a professor of climate and cryosphere research at the Institute of Meteorology and Geophysics, and Chairman of the Center for Climate and Cryosphere at the University of Innsbruck, Kaser delved into the intricacies of glacier mass and energy balance, climate fluctuations and their impact on glaciers, and the climatology and hydrology of tropical mountain regions. He also investigated the drivers of glacier mass changes, contributing largely to what we know about these complex systems today.

Kaser's involvement in the IPCC illustrates his status as one of the foremost authorities in the field, shaping global discourse on climate change through his research, publications, and advisory roles.

In 2010 the U.N. IPCC overestimated the number of unhealthy receding glaciers despite Georg Kaser's objection. Kaser stated, "This number (of receding glaciers reported by the IPCC) is not just a little bit wrong, but far out of any order of magnitude ... It is so wrong that it is not even worth discussing[101]."

In an article entitled "Ice Melts in Norway to Reveal Ancient Artifacts", one journal states,

> "It is not news that due to global warming; ice is melting adjacent to the poles at record speeds. Mountain ice and glaciers hold history to our planet's past, and this week archeologists published their findings. A treasure trove of artifacts circa 300-1500 AD, with peak activity in 1000 AD, were found in Scandinavia. In Central Norway, over 3,000 artifacts, including textiles, animal bones, hunting tools, and more were found nearly untouched. What is really cool about this find is that normally, Earth's elements would cause a breakdown in organic materials. And in the case of these glacial treasures, they have fought the good fight and won[102]."

The melting of glaciers predates the significant increase in CO_2 emissions, which began after World War II. With many glacial records dating back to at least 1850, we can see that many glaciers were receding even before then, driven by natural climate variations.

As you may remember from earlier in Part Four, Antarctica houses about 90% of the world's glacier ice, with Greenland holding roughly 9%, leaving the remaining 1% distributed across glaciers worldwide, providing perspective on their vast scale.

Observing glacier retreat served as a personal tipping point for me. It led me to recognize the Earth's inherent cycles—long-term and short-term—alongside abrupt changes influenced by uncontrollable events. Historical records abound with examples of temperature fluctuations, including winters devoid of cold and summers blanketed in snow.

My journey into understanding climate change began two decades ago with a seminar on man-made global warming. Faced with conflicting information, I delved deeper into the subject, seeking clarity. A pivotal

---

[101] https://www.climatedepot.com/2010/12/08/special-report-more-than-1000-international-scientists-dissent-over-manmade-global-warming-claims-challenge-un-ipcc-gore-2/
[102] https://www.discovery.com/nature/ice-melts-in-norway-to-reveal-ancient-artifacts

moment in solidifying my skepticism toward human-induced climate change occurred when I learned of a glacier in the French Alps that *expanded* throughout the 1800s, starting hundreds of years earlier during the Little Ice Age. After nearly engulfing a grand cathedral, a fervent prayer ceremony led to its retreat in 1851, a natural event likely reinforcing the faith of the local people. It has been receding ever since. This happened long before the ramp up of CO_2 beginning after World War II.

Glacier Bay, Alaska, provides another compelling illustration. Initially stretching to the ocean in the late 18th century, glaciers in this region have been in continuous retreat. Yet, proponents of climate alarmism struggle to explain natural warming trends following the Little Ice Age and blame all current warming on CO_2. This simply doesn't make sense. The majority of Glacier Bay melted before fossil fuels were in use.

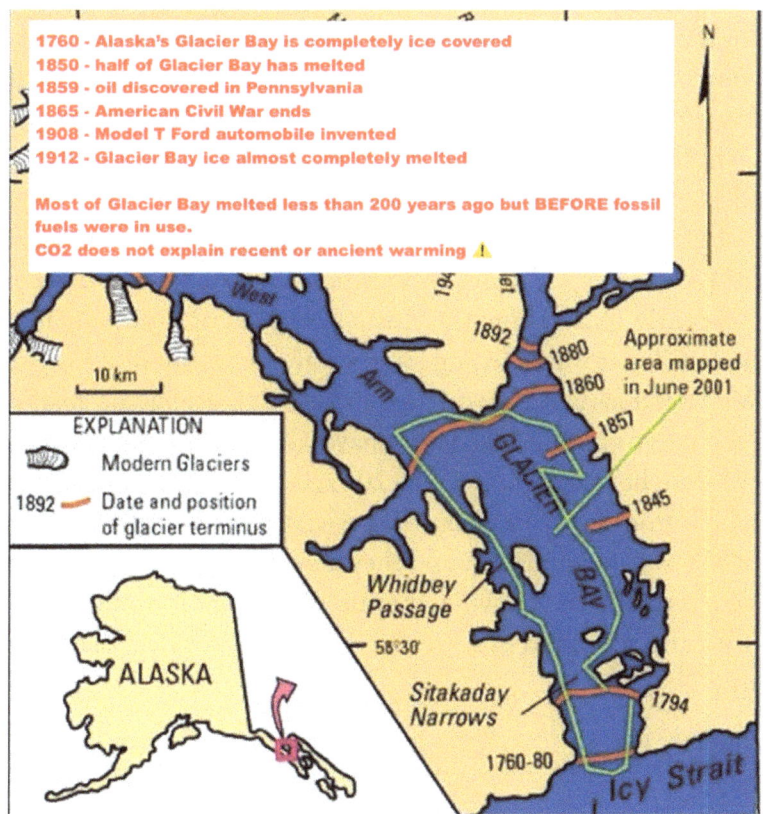

Source: USGS https://pubs.usgs.gov/of/2002/0391/images/GlacierBayTerminus.gif

The Earth's glacial history reveals a recurring pattern of glacier growth and retreat. Exposed tree stumps preserved for thousands of years beneath receding glaciers prove the point. They have even found 10,800-year-old stumps under the Alps. If they had been exposed to the elements, they would have decayed to soil. It was warmer when they grew in the first half of the warmer Holocene Climate Optimum; then it became colder.

There is the 5,300-year-old Otzi the "iceman" from the Alps who was frozen under the ice for all that time on the Austrian-Italian border. If he had been exposed, he would have been lost to the elements. Melting glaciers in Norway are revealing Viking and Roman artifacts. It is a race against time, because once they are exposed by the melt, they deteriorate very rapidly.

All of these discoveries point to warmer times in the early Holocene Climate Optimum, the Roman, and Medieval Warm Periods. If you do a little searching, you will see that we are finding things from these warming times from under the melting glacial ice. Glaciers have grown and receded throughout time. It had to be a lot warmer for a long period of time for forests to grow where glaciers are now. The shifting Arctic tree line, once positioned further north during warmer periods like the early Holocene era, underscores the planet's dynamic climate history.

The Jakobshavn glacier in Greenland, known to have receded since at least 1851, defied expectations by experiencing three consecutive years of growth recently, challenging simplistic narratives of linear decline[103]. The Jakobshavn Glacier was receding long before the addition of man-released CO_2. Why was it natural 150 years ago and now it is supposedly man-caused?

According to NASA, the Jakobshavn Glacier in Greenland grew for the third straight year from 2016 to 2019; it grew from 22 to 33 yards per year and has thickened. There is no current data on this glacier. So, my bet is that it has continued to grow and they don't want to talk about it. In each of the articles about the growing, they predicted that it would be shrinking soon. They also suggested that it is the cold Atlantic waters that are causing it to grow[104].

---

[103] https://climate.nasa.gov/news/2882/jakobshavn-glacier-grows-for-third-straight-year/
[104] https://climate.nasa.gov/news/2882/jakobshavn-glacier-grows-for-third-straight-year/

Chapter Fifteen: Glaciers & Natural Cycles 193

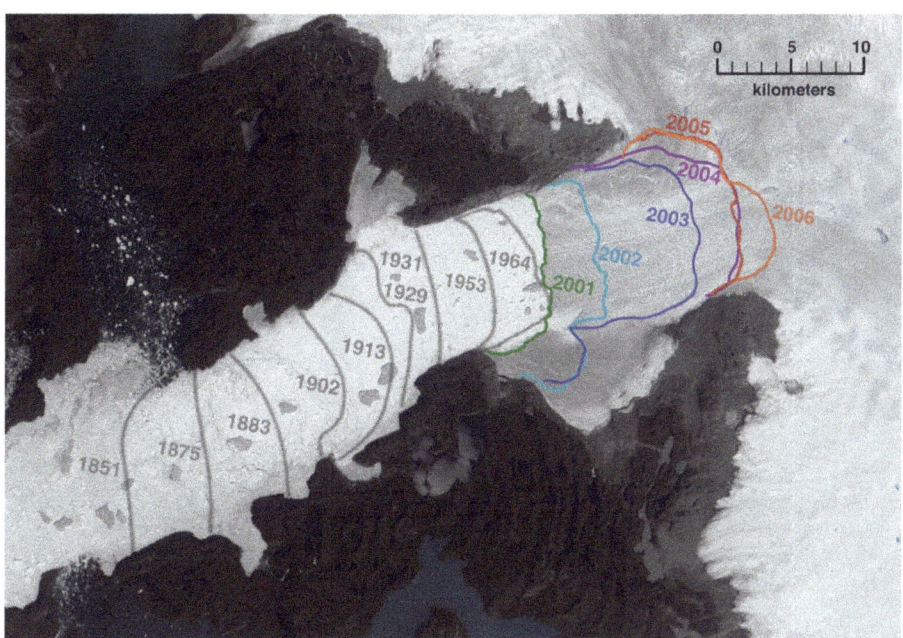

Source NASA Scientific Visualization Studio https://svs.gsfc.nasa.gov/3395

Some artifacts that have been frozen in glacial ice for millennia are emerging around the world. These artifacts deteriorate rapidly when exposed to air, meaning it is their first exposure. This also means it was warmer when they were frozen in place. They have remained frozen this entire time; there's no other option.

Of the items we have found in glaciers include the Iceman Otzi (mentioned above), an ancient arrow[105], an ancient forest[106], and other artifacts in a Swiss mountain pass that have been buried for at least 2,000 years[107].

In essence, the story of glaciers serves as a testament to the Earth's complex and ever-changing climate; we must approach climate narratives with discernment and understanding of larger natural cycles. Nature is in control, not us.

---

[105] https://www.nytimes.com/2023/09/23/climate/arrow-glacier-melting-norway.html
[106] https://www.huffpost.com/entry/mendenhall-glacier_n_3975699
[107] https://www.timesnownews.com/technology-science/melting-glaciers-uncover-a-swiss-pass-that-has-been-buried-for-at-least-2000-years-article-94197326?mc_cid=64bb2b0a99&mc_eid=02b86b93de

CHAPTER SIXTEEN:
MELTING GLACIER LIES

"It certainly does not follow logically that CO2 emissions drive a warming trend that began prior to widespread fossil fuel use and that has yet to reach the magnitude of the medieval warm period when Vikings colonized Greenland. Nor is a climate catastrophe implied by the presently observed rate of warming. Not only is the debate not over; it is expanding."
Dr. Charles Clough
Atmospheric Scientist, Chief of the Atmospheric Effects Team with the Department of the Army at Aberdeen Proving Ground 1982-2006

The narrative surrounding Greenland's ice melting and its purported contribution to rising sea levels is often sensationalized and misleading. There are no shortage of stock images of polar bears or penguins trapped on slivers of ice amidst a supposedly melting glacier landscape.

Greenland, an island cloaked in miles of ice and heavy snowfall, sees its ice cover replenished annually, despite hysteria around melting glaciers. This continuous addition of snow and ice to Greenland's mass contradicts the narrative propagated by mainstream media outlets.

Image Courtesy of Adobe (This is a composite picture. Polar bears live in the north and penguins in the south. They never meet in the wild.)

Although large puddles or shallow lakes of water may temporarily form during warm spells, they freeze over rapidly as temperatures drop, negating any significant long-term impact on Greenland's ice cover. The portrayal of these events as evidence of widespread ice loss is misleading. Contrary to alarmist claims, data from reliable sources, such as the polar portal managed by the Danish government, consistently demonstrate Greenland's net gain in ice mass. This contradicts the prevailing narrative of Greenland's ice melting catastrophically.

The PolarPortal.dk website clarifies the distinction between surface mass balance (SMB) and total mass balance, emphasizing that SMB accounts for the difference between snowfall and runoff. This shows that there is a surface mass gain every year between 180 and 600 gigatons (billion tons)[108].

---

[108] http://polarportal.dk/en/greenland/surface-conditions/

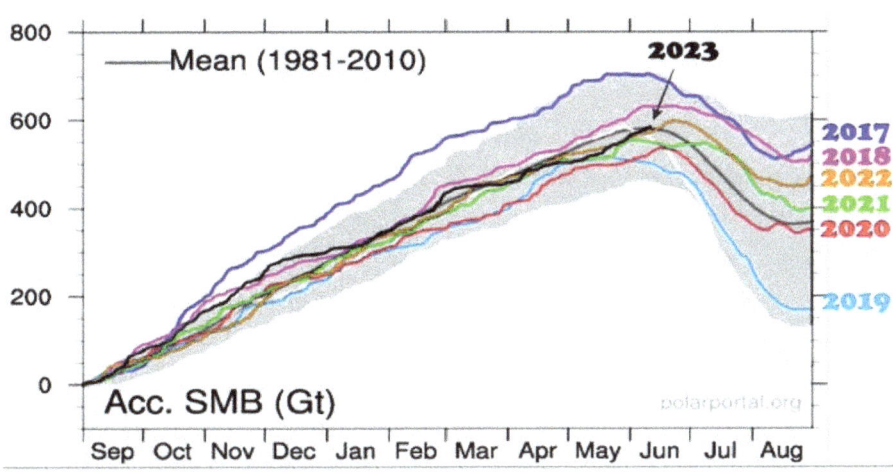

Source of the graph: http://polarportal.dk/en/greenland/surface-conditions/
Accessed 1/1/23

Despite the science, CNN posts headlines such as "Greenland Ice Melting on Tuesday Could Cover Florida in Two Inches of Water[109]..."

The amount of Greenland ice that melted on Tuesday could cover Florida in 2 inches of water

By Rachel Ramirez, CNN
3 minute read · Updated 12:15 AM EDT, Fri July 30, 2021

The article states, *"It would be enough to cover Florida in two inches (5 cm) of water...melting from Greenland is expected to raise global sea level between 2 and 10 centimeters by the end of the century..."*

Another propagandist article reads, *"Greenland is currently losing 234 billion tons of ice per year. That's enough ice to pack into 6,324 Empire State buildings. Ice is melting seven times faster now than it was in the 1990s. Greenland's accelerating rate of ice melt is one of many major changes in the region. The amount of sea ice in the Arctic Ocean has been decreasing"*[110]

---

[109] https://www.cnn.com › 2021 › 07 › 29 › us › Greenland-ice-melting-climate-change › index.html
[110] https://scied.ucar.edu/learning-zone/climate-change-impacts/greenlands-ice-melting

And finally, NASA writes in an article titled "Antarctica Melting Six Times Faster Than in the 1990's:

> "The peak loss coincided with several years of intense surface melting in Greenland, and last summer's Arctic heat wave means that 2019 will likely set a new record for polar ice sheet loss, but further analysis is needed. IPCC projections indicate the resulting sea level rise could put 400 million people at risk of annual coastal flooding[111]."

The following graph shows what's been happening with the total amount of glacier ice. It amounts to a rounding error, but don't let facts get in the way of a good story! In this graph below, you'll see two headlines: the one on the left states, "What the Media Shows You: Magnified Ice Loss Only: Greenland Cumulative Ice Mass Anomaly Since 1980," while the image on the right states, "What the Reality Is: Ice Loss vs. All Greenland Ice: Greenland Total Ice Mass Changes Since 1980."

On the left side, you see that there is a dramatic decrease in Cumulative Ice Mass, but that's because the chart essentially illustrates a single snowball in what constitutes an entire soccer field of ice. They're showing a single melting glacier while the reality is that when you take total ice mass into consideration, there have been no detectable changes in ice loss.

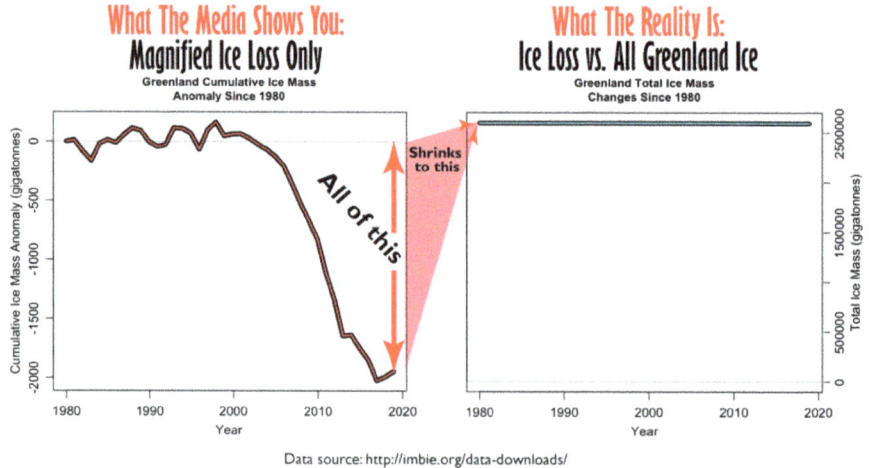

---

[111] https://climate.nasa.gov/news/2958/greenland-antarctica-melting-six-times-faster-than-in-the-1990s/

Even if it was true that Greenland was melting, it would take centuries to melt. On top of that, if it did melt, it wouldn't suddenly begin flooding Florida, or any other nation or land mass, for that matter. Greenland receives an incredible amount of snow every year, a fact that can be difficult to grasp. In 1988, World War II planes were discovered under 260 feet of ice, and in 2018, even more planes were found buried under 300 feet of ice. Each year, the accumulation of snow causes glaciers to grow.

When glaciers extend over the ocean or meet it, they break off and form icebergs, which act as floating fertilizer tablets for the ocean. Icebergs contain essential minerals that are lacking in many parts of the ocean. Many areas of the oceans have very little marine life due to the shortage of iron and other minerals. Introducing iron to the ocean prompts phytoplankton blooms. This increases primary productivity because phytoplankton are vital to the ocean's food chain, as we've covered.

Due to the constant influx of snow, glaciers continually expand, resulting in the formation of enormous icebergs, which is a natural occurrence. The Titanic famously collided with a massive iceberg in April 1912, sinking the vessel. Glaciers weren't a foreign object; the ship's captain and crew were well aware of the risk before they departed on that dreadful voyage.

The Greenland ice sheet didn't melt away in the warmer Holocene, Roman, and Medieval periods. There is no reason to believe that it will melt now.

The facts simply don't support the catastrophic climate change narrative.

POLAR BEARS ARE DOING JUST FINE – BETTER THAN FINE

The issue of polar ice caps and Arctic Sea ice elimination has been a subject of much debate and misinformation in climate discussions. Various narratives have been spun, often with climate alarmists cherry-picking data to fit the cult of climate change agenda. However, a closer examination of historical records and scientific evidence reveals a different truth.

Susan Crockford is a highly-experienced zoologist with over four decades of expertise in Arctic animals. Her research, particularly focusing on the Holocene history of Arctic fauna, has made significant contributions to our understanding of these ecosystems. Formerly serving as an adjunct professor at the University of Victoria, British Columbia, she now dedicates her efforts full-time to her private consulting firm, Pacific Identifications Inc.

In addition to her academic and professional pursuits, Crockford maintains a valuable blog https://polarbearscience.com/ that offers reliable and unbiased information about the Arctic and Antarctic regions. Her blog serves as an essential resource of accurate insights into polar bear ecology and related topics, devoid of any misleading or skewed interpretations. On the blog, Crockford states:

> *"Inuit in eastern Canada (Davis Strait) report a marked decline in ringed seal numbers since 1950, but otherwise there were no reports of population declines or reduced health in Arctic seals or walruses. In both the Arctic and Antarctic, less summer sea ice and increased primary productivity over the last two decades has meant more food for all animals and explains in part why polar wildlife has been thriving."*

The increase in polar bear populations contradicts the dramatized narrative of polar bear decline in mainstream media. In reality, polar bear populations have seen an upward trend, thanks in part to conservation efforts such as a hunting ban. Certain regions, like southern Greenland, have witnessed the adaptation of polar bears to changing environments, with some populations thriving in areas with reduced or little sea ice.

Increased primary production in the Arctic is leading to a higher abundance of prey such as seals, which are a primary food source for polar bears. Primary production is the bottom of the food chain, it provides the food for the next several steps up in the food chain. Plankton is the primary food source in much of the open ocean, much like plants provide food on land. This results in healthier seal populations, which positively impact polar bear populations by providing them with more food resources.

While it's true that Arctic Sea ice has decreased by about 20% since the 1980s, this change hasn't adversely affected polar bear populations. There have been periods in Earth's history when temperatures were warmer, and ice cover around the world was less extensive than it is now.

Polar bears have survived and adapted to various climatic conditions throughout history, including warmer periods such as the Medieval Warm Period, the Roman Climate Optimum, and the Holocene Climatic Optimum periods in Earth's history when warmer temperatures and less extensive ice cover were the case for a short season before freezing back up to what modern day climate alarmists consider "normal."

The graph below titled "Holocene Temperature Variations" depicts temperatures from 12,000 years ago to present. Between 8,000 and 4,000 years ago there was a period of warmer climate than today. This historical data challenges the narrative that current temperatures are unprecedented in Earth's history. This evidence has been suppressed or altered to fit the narrative of ongoing manmade global warming. Despite the warmer climates in previous eras, polar bears, have managed to survive.

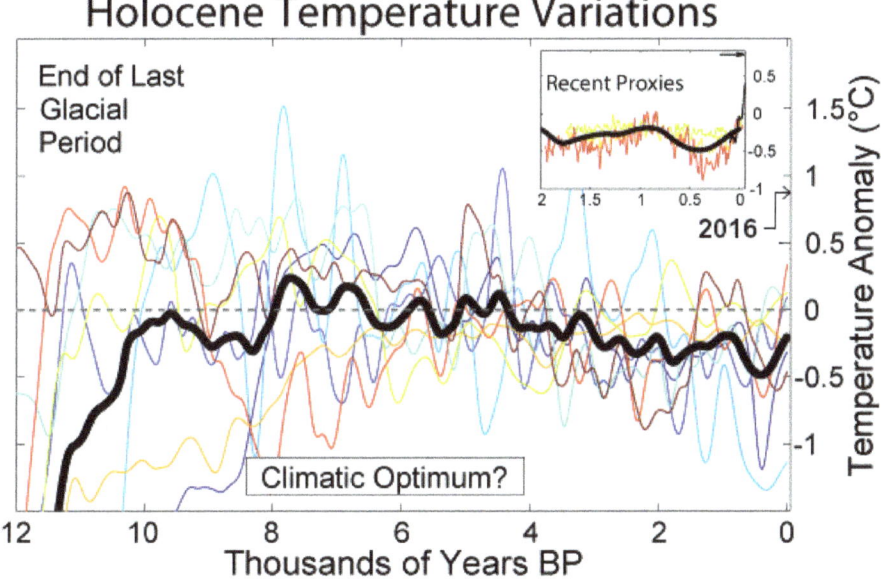

Source: Wikipedia Accessed 3/12/2024
https://en.wikipedia.org/wiki/Holocene_climatic_optimum

In fact, according to the *New York Times*, in the 1950's there were approximately 5,000 polar bears and in 2015 there were up to 31,000 of them. This has much to do with increased conservation efforts as well as more primary production, which means increased polar bear food.

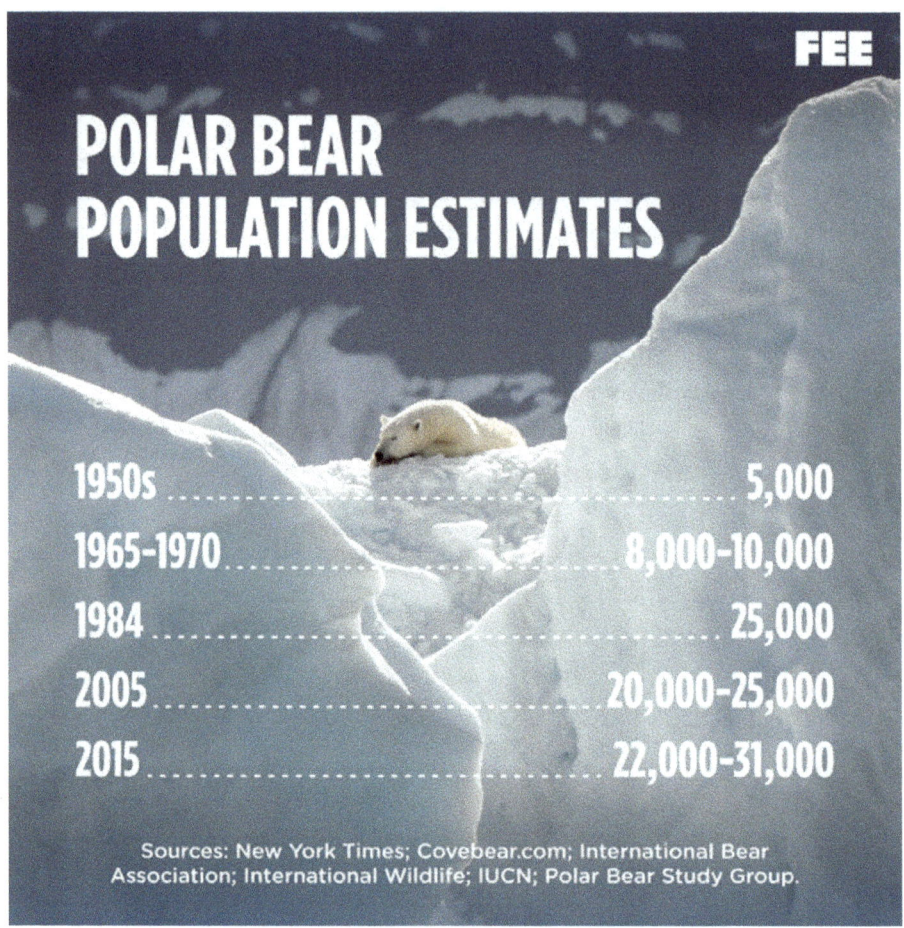

Susan Crockford, the polar bear specialist mentioned earlier in this chapter, faced backlash for reporting the truth about polar bear numbers. Her honesty led to her dismissal from the University of Toronto. The same type of censorship for speaking out against the mainstream narrative of climate change has also happened to many of the world's most respected and published doctors in history in fields such as cardiology and epidemiology during 2020-2021.

And, I, too, have been the victim of censorship for my stance on this topic; I've been banned from LinkedIn, and Facebook has restricted numerous factual posts that contradict the narrative. Sometimes people are censored because they're "flying too close to the sun" with regards to climate change and the fact-checking entities (mostly AI) flag these truth-tellers and censor or shadow ban their content.

For many academics and scientists, spreading misinformation from the cult of climate change and adhering strictly to the unscientific climate change doctrine gets them more funding and job security, while dissenting voices often face negative consequences.

Here are a handful of examples of polar bear fear-mongering propaganda meant to frighten you into accepting things that are not in your best interest:

1. Climate crisis: Polar bears face extinction by 2100 | World Economic Forum[112]

---

[112] https://www.weforum.org/agenda/2020/07/polar-bears-extinction-2100-new-study-climate-crisis/

2. Climate change: Polar bears could be lost by 2100 – BBC[113]

3. Global Warming Is Driving Polar Bears Toward Extinction, Researchers Say[114]

---

[113] https://www.bbc.com/news/science-environment-53474445
[114] https://www.nytimes.com/2020/07/20/climate/polar-bear-extinction.html

Polar bears are objectively cute; and who doesn't love a "teddy bear?" These are ways our heartstrings are pulled to buy into the climate lies and look no further than the image of an adorable polar bear standing in danger before we jump on the bandwagon.

In fact, perhaps polar bears don't need the arctic ice to survive at all, contrary to popular belief. In a recent end of the 2023 year summary on her website at polarbearscience.com, Crockford points out:

> *"Polar bears and sea ice fail to implode in 2023 as predicted. As this year draws to a close, it is worth noting that over the last 12 months — and contrary to predictions and headlines, including claims about 'the warmest year ever' — polar bears have not been reported dying, starving, or eating each other in large numbers, or relentlessly attacking people. On top of that, summer sea ice coverage in the Arctic has stalled for the last 17 years, not melted out in a death spiral of rotten ice[115]."*

A common tactic used by climate propagandists to manipulate data is to selectively choose a cold, snowy period as a starting point for analysis, effectively erasing the broader historical context of Arctic sea ice extent. For instance, the graph from the Department of Energy (DOE) below shows lower amounts of sea ice in the 1930s, 40s, and 50s, which is omitted from modern climate narratives[116].

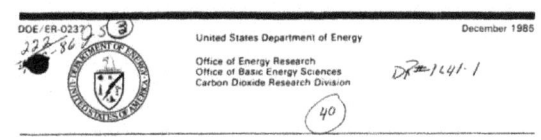

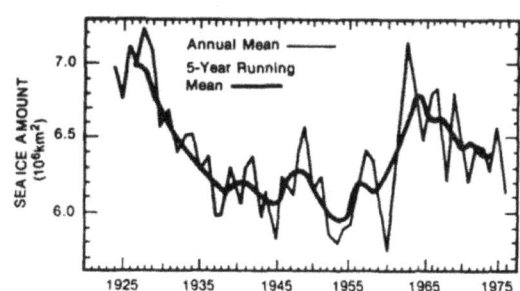

Figure 5.2. Annual mean and 5-year running mean sea ice amount in the Arctic Ocean from 1920-1975 (data from Vinnikov et al. [1980]).

---

[115] https://polarbearscience.com/2023/12/29/polar-bears-and-sea-ice-fail-to-implode-in-2023-as-predicted-with-special-thanks-for-your-support/
[116] https://archive.ipcc.ch/ipccreports/far/wg_I/ipcc_far_wg_I_chapter_07.pdf

The image below shows a *New York Times* article from October 19th, 1958. The article highlights historical climate reporting. It shows a submarine surfacing in the North Pole in an area with thinner and reduced sea ice compared to half a century ago, underscoring the physical changes occurring in the Arctic region. The article states:

> *"Some scientists estimate that the polar ice pack is 40 percent thinner and 12 percent less in area than it was a half-century ago, and that even within the lifetime of our children the Arctic Ocean may open, enabling ships to sail over the North Pole...Although the idea that a solid ice sheet covers the central Arctic has lingered stubbornly in the popular fancy, the northern cap of ice worn by our planet is actually a thin crust - on the whole, only about seven feet thick- over an ocean two miles deep in places."*

The decision to start satellite measurements in 1979, conveniently coinciding with a period of high ice extent, disregards earlier satellite data from the 1960s that contradicts the desired narrative. By selectively choosing data points and adjusting historical records, the true variability of Arctic sea ice extent over time is hidden. The modern version is an outright lie designed to mislead us.

Below is a graph of the polar sea ice from a 1990 IPCC report that illustrates clear data tampering. The charts show that there was significantly less ice before 1979, it was below average. In the IPCC 2001 graph on the lower half of the image, the lower quantity of sea ice was adjusted to a significantly higher than average amount of sea ice, which is blatantly false[117].

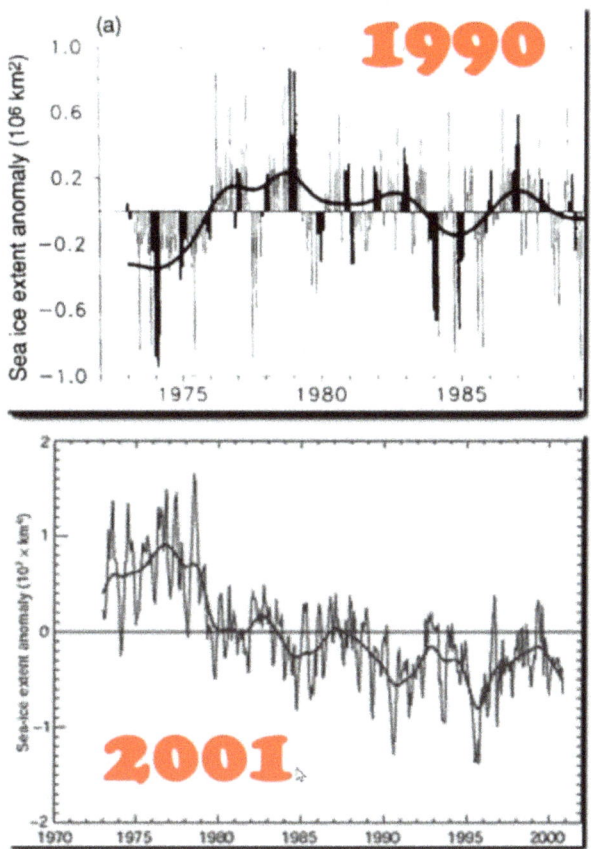

Source for the 1990 IPCC graph https://archive.ipcc.ch/ipccreports/far/wg_I/ipcc_far_wg_I_chapter_07.pdf
Source for the 2001 IPCC graph of arctic sea ice extent https://archive.ipcc.ch/ipccreports/tar/wg1/fig2-14.htm

I imagine it would be hard to sell the climate alarm narrative if the public had knowledge that the sea ice is increasing and less volatile today than in previous cycles with less CO_2.

It's essential to recognize that Arctic sea ice extent is cyclical, with natural fluctuations occurring over different time scales. While there has been a long-term slow decline in ice extent, historical records indicate periods of both

---

[117] https://realclimatescience.com/2022/12/lies-damned-lies-and-arctic-graphs/

greater and lesser ice extent than observed today. Additionally, the impact of floating ice on sea levels is minimal, as melting ice that is already in the ocean does not contribute to rising sea levels.

The graph below shows the annual sea patterns from 1995 to 2023. This graph isn't alarming. Yes, the trend is down some. At some point, we will see the ice returning in greater amounts because it is cyclical.

In the graph, titled "OSI Arctic Sea Ice Extent," Arctic sea ice varies from about 2.8 million square miles (4.5 million km) at its low point in September at the end of summer and 9.8 million square miles (15.8 million km) at its late winter maximum[118].

Furthermore, predictions of an ice-free Arctic have repeatedly failed to materialize. Dr. Peter Wadhams, a University of Cambridge Professor of Ocean Physics and Head of the Polar Ocean Physics Group in the Department of Applied Mathematics and Theoretical Physics at Cambridge is a notable climate alarmist who has made some particularly dramatic claims, all of which have failed to materialize.

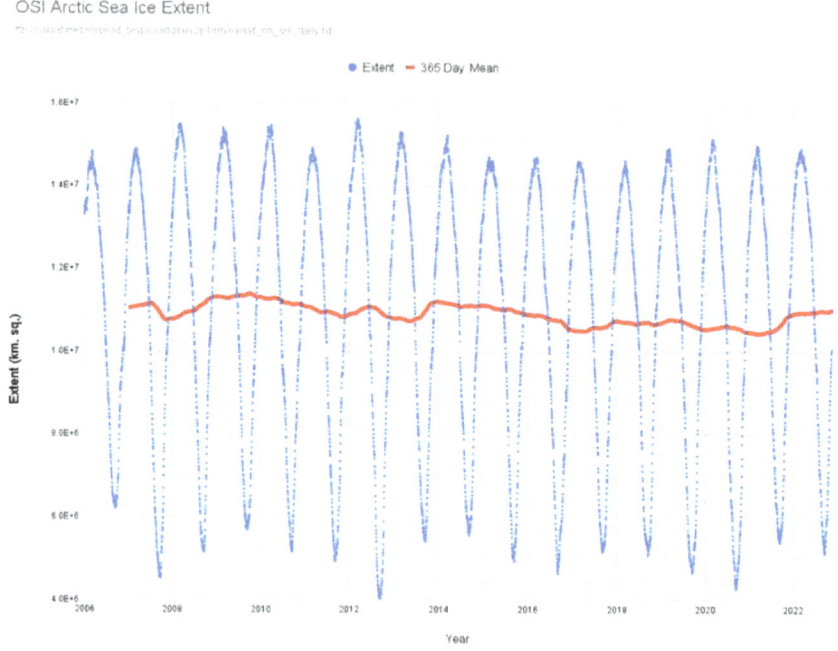

---

[118] https://nsidc.org/arcticseaicenews/charctic-interactive-sea-ice-graph/

Additional hysteria and inaccurate predictions regarding the demise of Arctic Sea Ice include one from *The Guardian* on September 17th, 2012. In this article Wadhams predicted a "global disaster" unfolding in the Arctic, warning about the "final collapse of the Arctic Ice in the summer months within four years."

That warning took place in 2016.

Wadhams also attacks methane, which is one of the tiniest, inconsequential of trace gases in our atmosphere. In another article from *The Guardian* on August 21st, 2016, Wadhams writes "The summer ice cover at the north pole is about to disappear, triggering even more global warming." He also said, "Next year or the year after, the Arctic will be free of ice." And in a 2020 YouTube interview he predicts an ice-free Arctic again[119].

In short, despite his credentials, Wadhams has made sensational predictions about the imminent collapse of Arctic sea ice, only to be proven wrong time and again. The failure of such predictions underscores the limitations of current climate models, and the audacity some "scientists" have to portray false information to the public without retribution.

Similar to Wadhams, James Hansen, the godfather of climate hysteria, is still making doom predictions that don't come true as well. Nevertheless, the propagandists in the media keep quoting him. In the article below from June 24, 2008 in *The Argus-Press*, a daily newspaper published in Michigan, we read, "Hansen, echoing work by other scientists, said that in five or 10 years the Arctic would be free of sea ice in the summer."

---

[119] https://www.youtube.com/watch?v=KxDg3pgbW9g

> The Argus-Press • Owosso, Michigan • Tues., June 24, 2008
>
> By SETH BORENSTEIN
> AP Science Writer
>
> ## NASA scientist: 'We're toast'
>
> "We see a tipping point occurring right before our eyes," Hansen told the AP before the luncheon. "The Arctic is the first tipping point and it's occurring exactly the way we said it would."
>
> Hansen, echoing work by other scientists, said that in five to 10 years, the Arctic will be free of sea ice in the summer.
>
> Longtime global warming skeptic Sen. James Inhofe, R-Okla., citing a recent poll, said in a statement, "Hansen, (former Vice President) Gore and the media have been trumpeting man-made climate doom since the 1980s. But Americans are not buying it."
>
> But Rep. Ed Markey, D-Mass., committee chairman, said, "Dr. Hansen was right. Twenty years later, we recognize him as a climate prophet."

From Google News: https://news.google.com/newspapers?nid=1988&dat=20080624&id=7mgiAAAAIBAJ&sjid=7qkFAAAAIBAJ&pg=5563,4123490

This article goes on to praise him: "Dr. Hansen was right. Twenty years later, we recognize him as a climate prophet." However, Hansen hasn't gotten *any* of the doom predictions even remotely right. Where are those fact checkers when you need them?

In other interviews, Hansen recommends that we put an end to fossil fuels. In later chapters, we will see why that is actually a terrible idea.

In short, the Arctic sea ice is doing fine. There are long term cycles of all things in nature. Predictions of global warming are highly speculative, with the majority of mainstream predictions turning out to be very wrong. Arctic sea ice is cyclical; it changes all the time and is meaningless to sea level. What's more, less ice means more food for the few Arctic animals that inhabit that region.

CHAPTER SEVENTEEN:
ANTARCTICA HAS HAD NO MELTING IN 70 YEARS OR MORE

"The Antarctic continent has not warmed in the last seven decades, despite a monotonic increase in the atmospheric concentration of greenhouse gasses."
Hansi A. Singh & Lorenzo M. Polvani

Searching for information on Antarctica's ice, such as whether it's melting or not, often leads to a misleading narrative that contradicts reality. This distortion resembles the upside-down world depicted in the popular TV series "Stranger Things," where sinister forces emerge from an alternate reality beneath our own—an analogy fitting for the realm of climate propaganda and the energy policies it is driving.

Google is complicit in suppressing dissenting views, burying that information deep in search results, erasing it altogether, or making it impossible to find. The U.N.'s collaboration with Google ensures that alarmist narratives dominate search results, while dissenting voices are marginalized. Google's ownership of YouTube further exacerbates this issue, with censorship and biased search manipulation prevalent. These "search engines" often label pro-climate realism content as false despite its reputable sources.

Contrary to popular belief, Antarctica—a continent comparable in size to North America—has not experienced warming for 70 years, as indicated by a peer-reviewed study published in *Nature*[120]. Temperature records dating back to the 1950s corroborate this[121].

Another report entitled "Antarctica Setting Cold Temperature Records" reads, "Antarctica has 43 permanent, 44 seasonal, and dozens of automatic research stations operated solely or jointly by 42 nations, and many of those stations have measured record or near record cold temperature in the fall and winter of 2023[122]."

Antarctica has set new coldest ever records in the last 2 years. Surprisingly, mainstream media reported these, and other sources did as well[123,124,125].

Additionally, although Antarctic sea ice exhibits variability, it has generally been stable over time. Antarctica is noteworthy to the conversation of warming and cooling due to its extreme coldness, with temperatures consistently well below zero punctuated by occasional warm spikes. Propagandists seize upon these brief anomalies to perpetuate the false narrative of Antarctic melting, despite overarching evidence to the contrary.

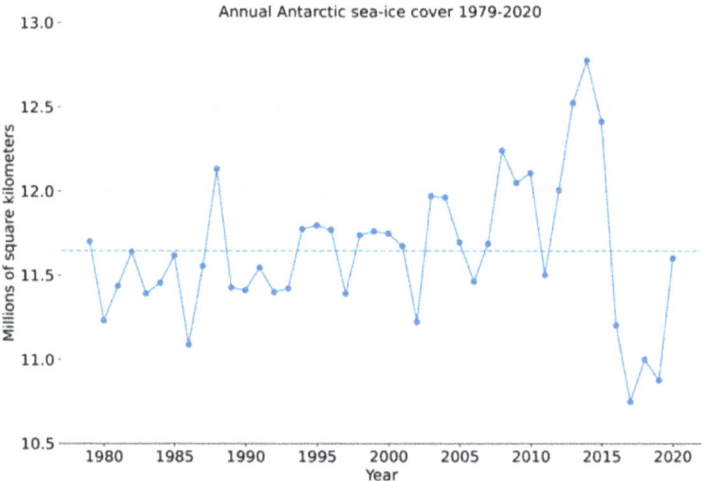

https://realclimatescience.com/2019/09/no-change-in-arctic-sea-ice-extent-for-almost-14-years/#gsc.tab=0

---

[120] https://www.nature.com/articles/s41612-020-00143-w
[121] https://www.nature.com/articles/s41612-020-00143-w
[122] https://coldweatherreport.com/2023/08/21/antarctica-setting-cold-temperature-records/
[123] https://coldweatherreport.com/2023/08/21/antarctica-setting-cold-temperature-records/
[124] https://www.cnn.com/2021/10/09/weather/weather-record-cold-antarctica-climate-change/index.html
[125] https://climaterealism.com/2023/11/new-antarctic-all-time-cold-record-flies-in-the-face-of-media-reporting/

Glaciers thrive on snowfall, which compacts under pressure to form ice. This phenomenon is crucial for sustaining glaciers, ensuring their growth and stability. Antarctica gets little snow. It is extremely cold, and the prevailing winds do not bring it much moisture to fall as snow.

The Antarctic Ocean, a recently designated body of water, serves as a vital habitat for krill, a keystone species in the marine food web. Harvested for food and omega-3 supplements, krill underscores the ecological significance of Antarctica.

Contrary to climate change predictions, Antarctica's cold summers and recent cooling trends challenge the climate change notion that both poles should warm uniformly. The region's low solar energy absorption maintains its perpetual coldness. Both poles release significant amounts of the Earth's heat to the far colder space that surrounds our planet.

While parts of Antarctica, notably the Thwaites Peninsula in western Antarctica, experience warming from more than 138 active undersea volcanoes, this localized phenomenon does not signify global warming. Media sensationalism often portrays this as the "doomsday glacier," exaggerating its potential impact on sea levels.

Heat maps of Antarctica depict localized warming around the Thwaites Peninsula, while the vast majority of the continent remains cooler than historical averages, as acknowledged by NOAA.

Graphs circulating on social media, purportedly illustrating declining sea ice in both the Arctic and Antarctica, often employ selective data starting from 1980. This cherry-picked approach obscures earlier evidence and misrepresents long-term trends.

NASA's Earth Observatory provides a more comprehensive view, revealing minimal fluctuations in Antarctic sea ice relative to the 1980–2010 average as is seen in the graph below entitled "Antarctic Daily Sea Ice Extent." (Remember that when it is winter in the north, it is summer in the south, and vice versa.)

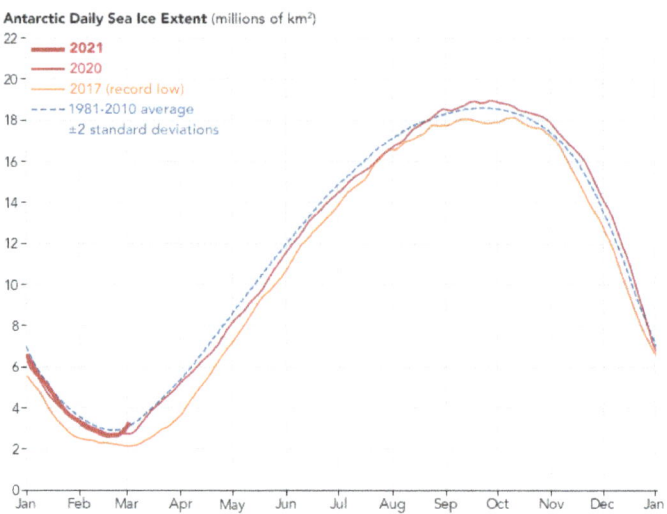

Photo Courtesy of NASA[126]

The next image shows that there was a loss of 2,500 gigatons out of 29 million gigatons in the graph entitled "Change in Total Ice Mass Antarctica, 1992-2017[127]." That's a loss of only 0.0086 percent which is a quantity that could be equated to a "rounding error."

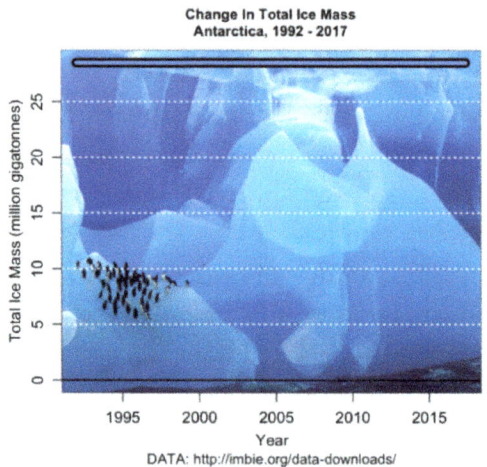

Photo Courtesy of http://imbie.org/data-downloads/

Antarctica and its ice are doing just fine; please help spread the word.

---

[126] https://earthobservatory.nasa.gov/world-of-change/sea-ice-antarctic
[127] http://imbie.org/data-downloads/

CHAPTER EIGHTEEN:
HOW EARTH'S WEATHER IS MADE

"Nature, not man, rules the climate system."
Joe Bastardi

Joe Bastardi holds a bachelor's degree in Meteorology and gained experience working at AccuWeather before joining WeatherBELL Analytics, a meteorological consulting firm. As a professional meteorologist and weather forecaster, Bastardi has authored several books on climate, including, *The Climate Chronicles: Inconvenient Revelations You Won't Hear from Al Gore* among others.

Known for his accurate weather predictions, he earns his living in the private sector by consistently forecasting future weather events.

Consider for a moment how the Earth's weather is formed: the sun is the primary source of energy for our planet, heating the equatorial regions around the equator more than the poles. Solar warmth causes water vapor, the most abundant greenhouse gas to evaporate, mainly from oceans. Solar warmth also causes water to evaporate from lakes, rivers, and agricultural irrigation (although in much smaller quantities than from the ocean.)

Evaporation is a cooling process that requires energy to convert liquid water into vapor, much like how our bodies sweat to cool off. The Earth is in "hot" pursuit of achieving temperature equilibrium, although that's not possible. The atmospheric circulation on Earth revolves around energy imbalance and the continual recalibration between the equator and the poles. The Earth attempts to get the same temperature everywhere; its water, winds, and air are always moving, striving for this unachievable equilibrium, as well as the Earth's daily rotation, which keeps everything in perpetual motion.

Warm, moist air is lighter than dry air, which is a counterintuitive. (Naturally, you'd think that moist air is heavier because of its water content.) When moist air rises from sun-warmed oceans toward colder regions, you can see how the Earth is seeking to homogenize temperatures across the globe.

Essentially, evaporation of ocean water, the ascent of warm, moist air, and the moist air's transport toward colder regions all work together to form the foundation of all weather phenomena on Earth.

The following image from NOAA shows Hadley cells, which show the circulation of the rising warm air from the oceans of the equator, warmed by the sun.

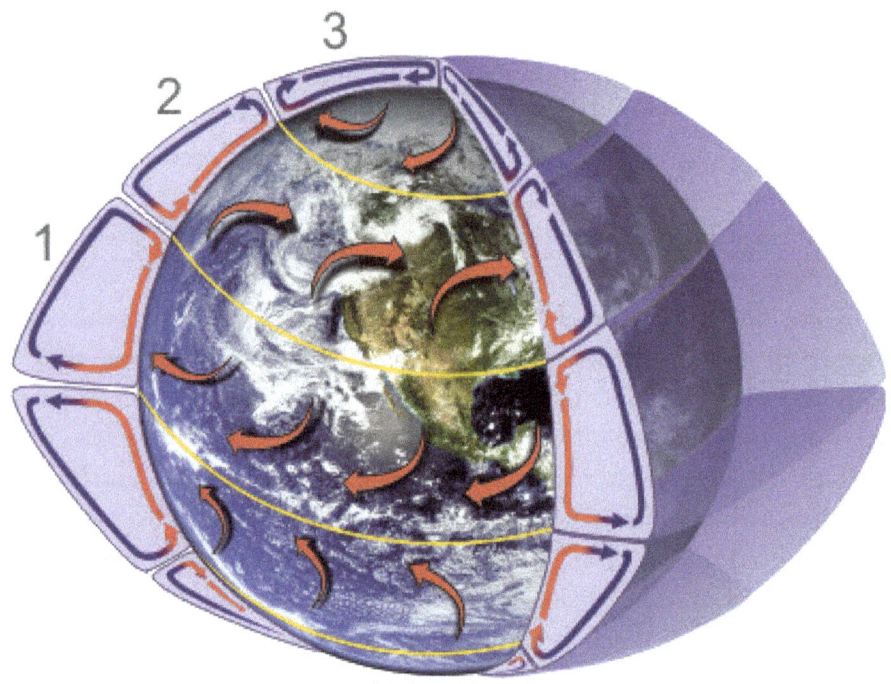

Source NOAA https://www.noaa.gov/jetstream/global/global-atmospheric-circulations

Warm air is always attracted to cold areas. The second law of thermodynamics tells us that heat always flows from hotter to colder regions. The greater the difference in temperature when they mix, the more likely we will have extreme weather events like tornadoes, hurricanes, and large storms. Because the Arctic area is warming more in the winter, there is a bit less difference between the warm air from the middle Ferrel cell and Arctic cell; and there is a little less difference than before between the tropical Hadley cell and mid-latitude Ferrel cells.

WEATHER & THE CO_2 LIE

In oceans with high evaporation rates, which serve as cooling mechanisms, the rising saturated air can result in tropical storms or hurricanes, depending on wind patterns and atmospheric pressure systems. They can be severe or mild.

Warm ocean temperatures fuel hurricanes and tropical cyclones, with the sun being the primary heat source for ocean warming. Furthermore, emerging research suggests heat from Earth's inner layers contribute to ocean warming from below.

The Earth's rotation produces winds that transport air of varying temperatures and moisture levels worldwide, causing it to mix and circulate, thereby influencing local weather patterns that are constantly changing.

As you can see, this system is far too complex with thousands of moving parts to be boiled down to a single trace gas, CO_2 as the "control knob of the temperature." That is just plain silly. Yet, that is what we are told over and over again.

HURRICANES

> *"There have been no detectable trends in hurricanes for over 100 years. There have been ups and downs of course, but over a long term, not much... There is a hint that the fraction of hurricanes that are the strongest has increased in the last 30 years, but then there are other papers that came out subsequently that said, 'No, no, it's just a return to natural variability*[128]*.'"*
> Steven Koonin

Steven Koonin is an accomplished American theoretical physicist who has held several prominent positions throughout his career. He served as the Director of the Center for Urban Science and Progress at New York University and is currently a professor in the Department of Civil and Urban Engineering at NYU's Tandon School of Engineering. Notably, from 2004 to 2009, Koonin held the position of Chief Scientist for the oil and gas company BP.

Following this, he served as the Under Secretary for Science at the Department of Energy during the Obama Administration from 2009 to 2011. Koonin is also the author of the insightful and candid climate book titled *Unsettled*.

The news media often portrays hurricanes as becoming more frequent, powerful, and water laden. However, historical data doesn't support these claims, as Koonin confirmed.

Hurricanes have long been a natural occurrence, with some of the deadliest hurricanes recorded during the 1770s and 1780s. The Great Hurricane of 1780, which devastated regions like Antigua and the Bahamas, claimed more than 20,000 lives, and wreaked havoc on settlements and ships.

An interesting and little-known sidebar: hurricanes helped win the Revolutionary War. The British had the greatest navy on Earth at the time. However, the Americans had help from nature while fighting the British

---
[128] https://jaxtoday.org/2022/01/25/controversial-author-steven-koonin-to-speak-about-climate-change-in-jacksonville/

during the Revolutionary War; a strong hurricane season wrecked many ships and sent many sailors to the bottom of the seas.

Analyzing historical records, it's evident that major landfalling hurricanes have fluctuated over the years, with no clear increasing trend. In fact, some decades have seen more major hurricanes than others, suggesting no consistent rise in their occurrence. In the past decades, we have been blessed with no landfalling major hurricanes at all.

As you can see in the graph below entitled "U.S. Landfalling Major Hurricanes," there have been fluctuations in hurricanes, but since the mid-nineteenth century, even with fluctuations, the occurrence of hurricanes remains fairly steady and almost even predictable. After a tumultuous hurricane season in the mid 2000s, major hurricanes ceased for some time.

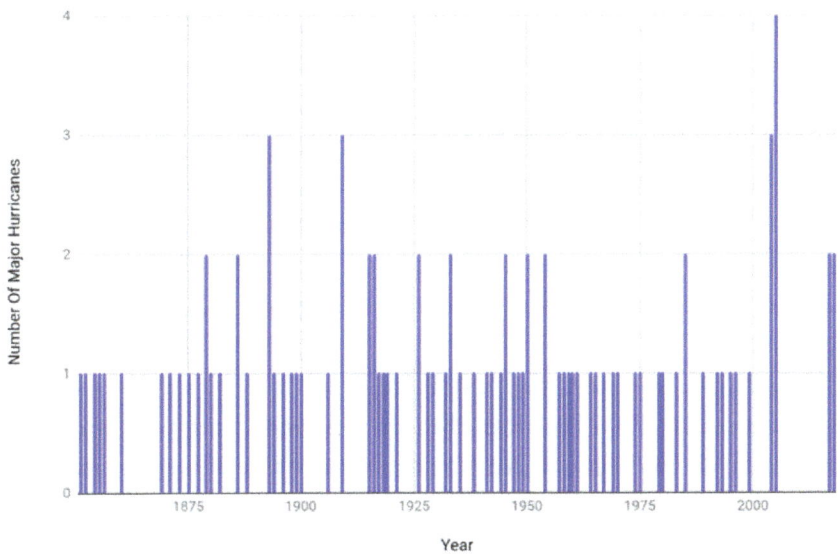

https://docs.google.com/spreadsheets/d/1UGtvUMpS4Rx5iqjgDWkVxwvljLj20pJX2dYkiZOhQek/edit#gid=790674289

On the graph below entitled "Global Major Hurricane Frequency," you'll see that hurricanes and tropical cyclones show no significant upward trend in frequency or intensity since 1980. Despite occasional fluctuations, there is no sustained increase in the overall activity of these weather phenomena[129].

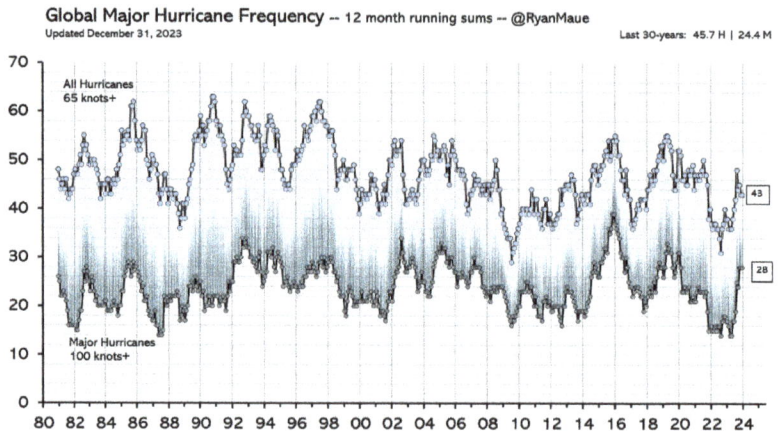

Source https://climatlas.com/tropical/

One common counter-argument to the clear and evident science is the increase of insurance claims for hurricane-related property damage. However, these arguments often fail to consider factors like inflation and economic growth, which often inflate the monetary value of losses. When adjusting for these factors, normalized costs reveal that hurricane damage as a percentage of the economy remains relatively stable or even declining over time.

In the chart below entitled "No Increase in Cost of Hurricane Damage," you'll see that the 2004 Hurricane Season incurred a bump in damage costs, but damage claims began decreasing again since then; the increase was not sustained over the past twenty years.

---

[129] https://climatlas.com/tropical/

Chapter Eighteen: How Earth's Weather is Made

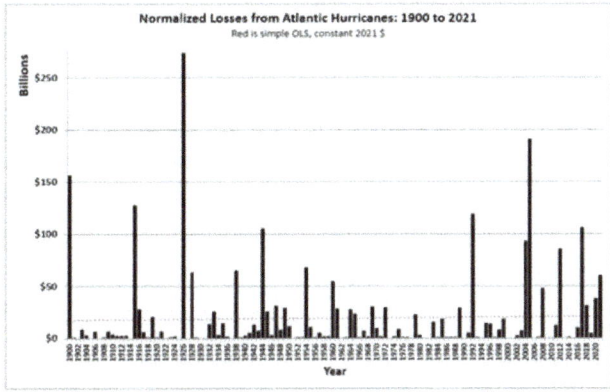

Source https://repository.library.noaa.gov/view/noaa/27147

While the media may sensationalize reports of billion-dollar weather losses, the reality is that advancements in building techniques and infrastructure have mitigated the impact of hurricanes on property. Additionally, the increase in population and property value over time naturally leads to higher absolute losses during hurricane events, but as a percentage of the total economy, these losses remain relatively consistent or have decreased.

In the following picture, you'll see an image of Miami, Florida, illustrating that in just under 100 years from 1925 to 2017, the population in Miami Dade County grew from about 100,000 to 2.7 million people. The county went from what appeared to be farmland and undeveloped land to a bustling city with massive high-rises, homes, and apartments for millions of people.

Cities all over the world have grown in this way and we are more prosperous. This means there is more to lose during a storm; naturally insurance claims are higher when there is more infrastructure that is vulnerable to destruction.

In the next image entitled "U.S. Billion-Dollar Hurricanes" you see that "billion-dollar hurricanes" have steadily increased since 1900; but that is also commensurate with billions of dollars of infrastructure having been erected and existing to be damaged in the first place. The creator of the image notes:

> *"This figure shows the steady increase in counts of 'billion-dollar' hurricanes with losses adjusted only for inflation. NOAA (mis)uses this methodology to claim that climate change is worsening U.S. disasters - which is catnip for many journalists and editors. All this graph actually shows is that hurricanes cause more damage because there is more wealth to be destroyed. If you want to look at climate trends, look at climate data, not economic data*[130]*."*

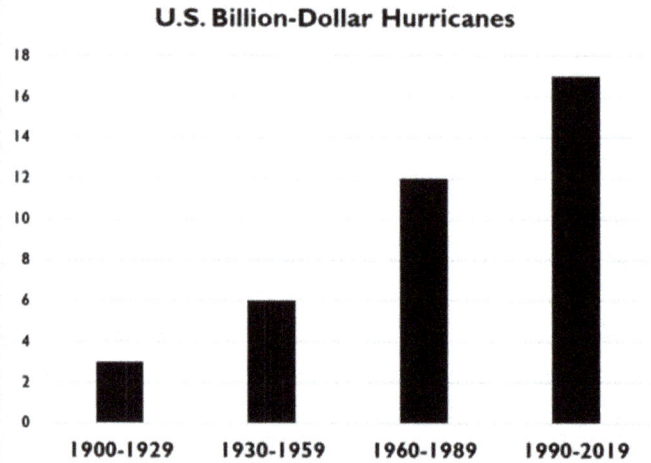

This figure shows the steady increase in counts of "billion-dollar" hurricanes with losses adjusted only for inflation. NOAA (mis)uses this methodology to claim that climate change is worsening U.S. disasters — which is catnip for many journalists and editors. All this graph actually shows is that hurricanes cause more damage because there is more wealth to be destroyed. If you want to look at climate trends, look at climate data, not economic data.

Photo Courtesy of https://rogerpielkejr.substack.com/p/pielkes-weekly-memo-19

---

[130] https://rogerpielkejr.substack.com/p/pielkes-weekly-memo-19

Take the Miami area that has grown to 2.7 million people with house and high-rise condo values in the $400,000 to $700,000 range. The value of the damage from a hurricane hitting Miami now would be a lot pricier than a hurricane strike 30 or 50 years ago. There are more people and more developed property to lose. Thirty years ago, there weren't multiple computers and a TV in nearly every room like there are now. All of these things increase the losses of any climate disaster. We have more to lose because of our fossil fuel derived prosperity.

NORMALIZED U.S. HURRICANE DAMAGE FROM 1900–2022

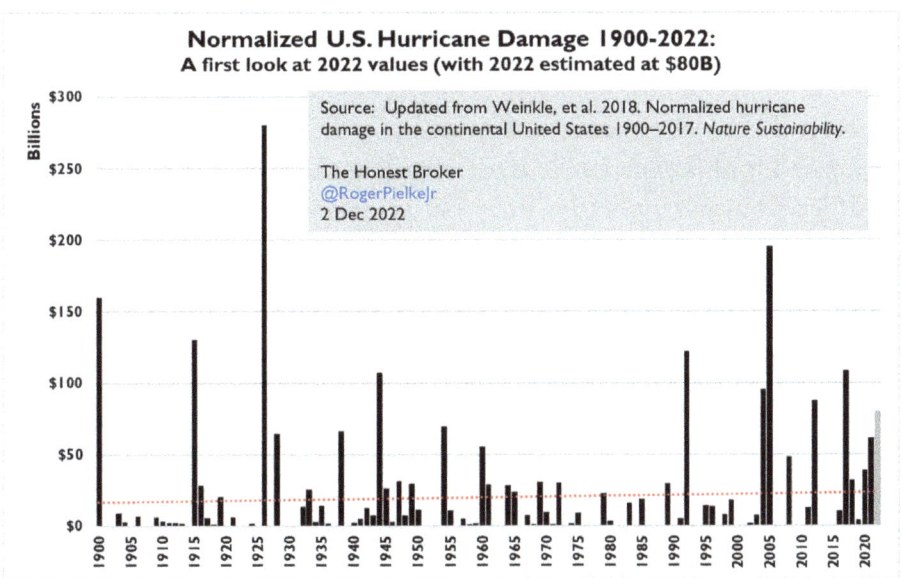

Photo courtesy of https://rogerpielkejr.substack.com/p/pielkes-weekly-memo-19

The graph above shows a first estimate of normalized U.S. hurricane damage from 1900 to 2022, with 2022 estimated at $80 billion (at 2x insured losses[131]). That estimate may yet be revised up or down. Of note, for the first

---
[131] https://www.bmsgroup.com/news/bmstropicalupdate12012022

time since we have presented this data, the trend since 1900 has turned positive, but only slightly so. Trends in U.S. hurricane landfalls — overall and major — remain slightly negative over 1900 to 2022.

It's critical to evaluate the data and narratives of the climate propagandists. They cherry-pick information to promote alarmism rather than provide accurate depictions of historical trends and future projections. The evidence herein shows that there is no compelling reason to anticipate a significant increase in hurricanes and tropical storms in the foreseeable future.

TORNADOES

"However, an examination of trends within sub-periods of the dataset is suggestive that some part of the long-term decrease in losses may have a component related to actual changes in tornado behavior[132]."
Roger A. Pielke Jr.

Roger A. Pielke Jr. is an esteemed American political scientist and professor, renowned for his expertise in understanding the politicization of science and decision-making under uncertainty. He served as Director of the Environmental Studies Program and was a Fellow of the Cooperative Institute for Research in Environmental Sciences (CIRES) at the University of Colorado Boulder from 2001 to 2007. Additionally, Pielke was a visiting scholar at Oxford University's Saïd Business School during the 2007–2008 academic year.

Pielke's diverse research interests include policy education for scientists in fields such as climate change, disaster mitigation, and world trade. He is a prolific writer, known for providing solid, honest information on climate and weather through his Substack. Pielke's contributions to understanding the intersection of science, politics, and policy have earned him widespread respect and recognition in academic and public spheres.

The trend for tornadoes, both in the United States and globally, is declining. Despite this, propagandists often claim that tornadoes are

---

[132] https://rogerpielkejr.substack.com/p/us-tornado-damage-1950-to-2021

becoming more common whenever one hits. Tornadoes naturally occur worldwide, but they generally aren't as severe as those in the U.S.

One reason for the perception of increasing tornadoes is the fact that we have improved detection methods, particularly with the advent of Doppler Radar in the early 1990s. With radar technology, more tornadoes are detected compared to the pre-radar era when sightings had to be reported by eyewitnesses. It's likely that many tornadoes causing minimal or no damage went unreported in the past. Now they are identified by Doppler Radar.

In the image below from NOAA.gov, you'll see that tornadoes are presented to be increasing since the 1950s when the graph commences. However, data like this can be misconstrued and manipulated, as we learned about in Chapter Eleven of this book. There should be a note with the graph describing how radar increases the reported tornadoes. Otherwise, it is like comparing apples to oranges.

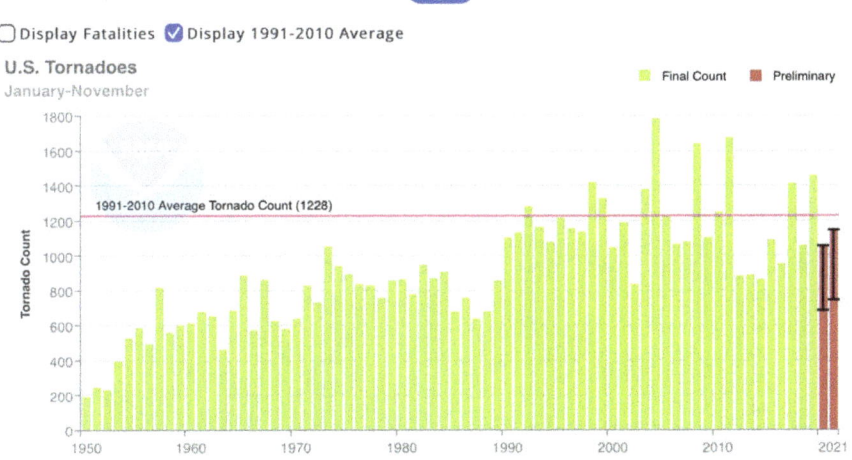

Photo Courtesy of NCDC.NOAA.Gov

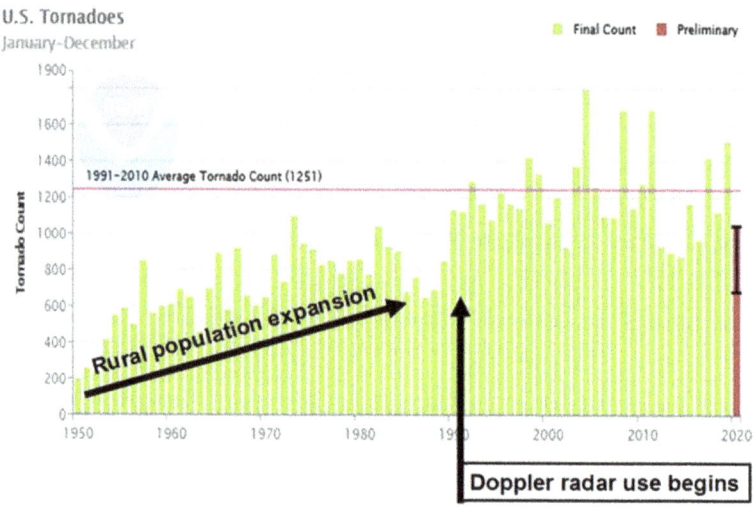

Photo Courtesy of NCDC.NOAA.Gov

In the next image entitled "Violent Tornadoes (EF3+) & CO_2," you'll see that violent tornadoes since 1950 are actually decreasing, even though CO_2 is rising. Of course, due to CO_2 being weaponized for climate lies, it's important to see accurate readings of the number of tornadoes next to CO_2 in the atmosphere. Truthfully, the relationship is not causal, but inversely correlated.

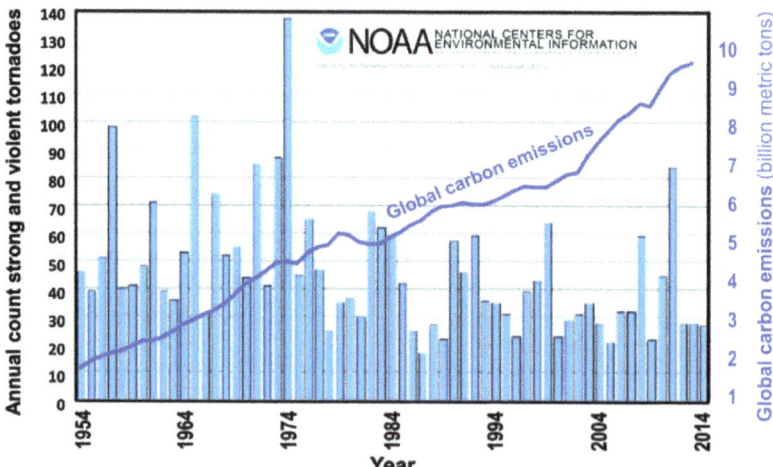

Graph courtesy of https://www.climatedepot.com/2021/12/12/watch-december-tornadoes-not-rare-tornadoes-used-to-be-blamed-on-global-cooling-in-1970s-tony-heller-of-real-climate-science-explains/
Source for raw data and graph https://www.ncei.noaa.gov/access/monitoring/tornadoes/

We must critically evaluate claims about tornado trends and recognize that the IPCC acknowledges the uncertainty in attributing trends in extreme weather events like tornadoes. This is what the IPCC said:

- "...observational trends in tornadoes, hail, and lightning associated with severe convective storms are not robustly detected;"
- "...attribution of certain classes of extreme weather (e.g., tornadoes) is beyond current modeling and theoretical capabilities;"
- "...how tornadoes or hail will change is an open question"

The downward trend in tornado activity contradicts the narrative pushed by climate propagandists, and it's crucial to rely on reliable data rather than sensationalized claims.

DROUGHTS

"We know that carbon dioxide has been a much larger fraction of the earth's atmosphere than it is today, and the geological record shows that life flourished on land and in the oceans during those times. The incredible list of supposed horrors that increasing carbon dioxide will bring the world is pure belief disguised as science.[133]*"*
Harrison Hagan Schmitt

Harrison Hagan Schmitt is an American geologist, whose diverse career has spanned academia, space exploration, and politics. He served as a professor of Engineering at the University of Wisconsin-Madison and became an Apollo 17 astronaut. Schmitt further distinguished himself as a former U.S. Senator of New Mexico.

As part of the Apollo 17 mission, Schmitt became the only person without a background in military aviation to walk on the Moon. His achievements in both science and public service have solidified his legacy as a pioneering figure in American history.

---

[133] https://www.wsj.com/articles/SB10001424127887323528404578452483656067190

Droughts have been a part of Earth's history since ancient times, thousands of years ago. Despite claims to the contrary, data from the past few decades shows a decline in the extent of global droughts.

The notion that droughts are becoming more common is contradicted by historical evidence, including records from the Bible and other sources documenting droughts throughout the world over various periods. In fact, data spanning from 1982 to 2012 indicates a decreasing trend in global drought occurrence.

Even in Europe, records since 1870 show no increase in drought frequency, as illustrated by graphs depicting wet and dry years in the region.

The occurrence of major drought events in history, such as the Dust Bowl in the 1930s and the Medieval megadroughts in North America, predates the industrial era and the significant rise in CO_2 levels. This challenges the narrative that higher CO_2 concentrations lead to more frequent droughts.

Mr. Schmitt agrees with the notion, stating, "Crop yields in recent dry years were less affected by drought than crops of the dust-bowl droughts of the 1930's, when there was less carbon dioxide. Nowadays, in an age of rising population and scarcity of food and water in some regions, it's a wonder that humanitarians aren't clamoring for more atmospheric carbon dioxide. Instead, some are denouncing it[134]."

The following graph shows the percentage of the world in drought and the severity. Entitled, "Global Integrated Drought Monitoring and Prediction System" by Nature.com, you can see that since 1997, the percentage of the globe in drought has decreased slightly, but perceptibly enough to the naked eye on this graph. According to this graph, you see that approximately less than 0.1% of the world is under severe to extreme drought at any point in time between 1982–2012. These figures as reported by Nature.com originally come from Standardized Precipitation Index Data via MERRA-Land.

---

[134] https://www.wsj.com/articles/SB10001424127887323528404578452483656067190

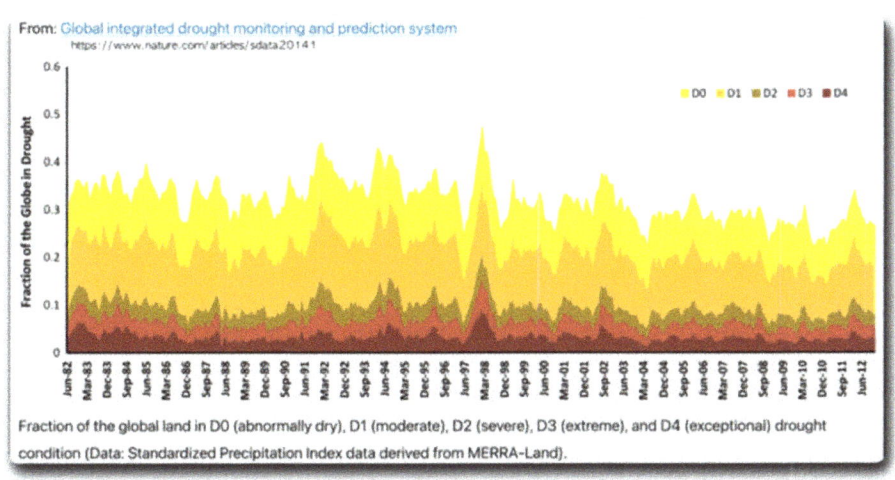

The next image provided by Karl Kahler of the Bay Area News Group originally from data provided by E.R. Cook et al, Earth-Science Reviews states: "Evidence from tree rings shows that drought was historically much more widespread in the American West than now, while the 20th century was wetter than normal. The small increase in 'drier' weather on this graph starting just before the year 2000 is measurably less significant than in medieval megadroughts."

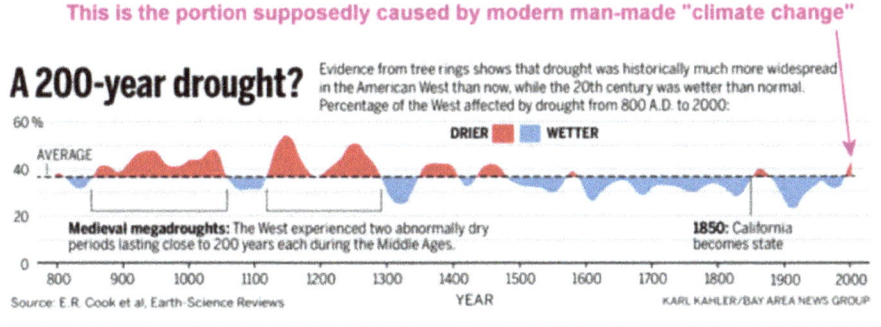

Although the graph above is used to promote climate alarmism and make it seem like we're heading into another "megadrought," data from NOAA, such as the "Packers graph" below suggests no significant trend in drought occurrence in the U.S. overall.

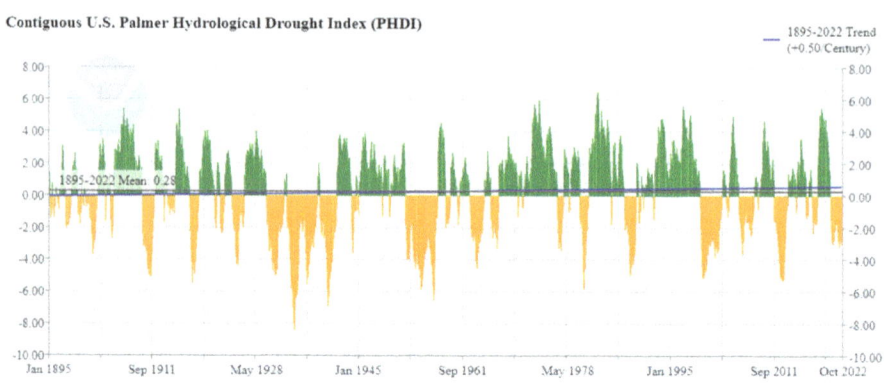

While California has experienced multi-year droughts, recent trends show increased rainfall leading to full reservoirs despite previous drought concerns. Similarly, Texas has seen fluctuations in drought patterns, but overall, there is no consistent trend indicating more frequent droughts.

Contrary to popular belief, damages from floods worldwide have decreased over time, further challenging claims of worsening weather-related disasters.

The rise in atmospheric CO_2 levels has benefits, as you are now well aware, including making plants more drought-resistant. This is due to reduced water loss through the stomata, which are minute pores in the epidermis of the leaf or stem of a plant, forming a slit of variable width which allows movement of gases in and out of the intercellular spaces.

This demonstrates the complexity of climate dynamics and the need to critically evaluate claims made by climate propagandists. Plants tolerate heat and droughts better with more CO_2 …and three to five times more CO_2 would confer maximal benefits. Our world is improving and the plant world is doing better because of more CO_2. We aren't adding it fast enough, actually. Let others know the benefits of CO_2. And that floods and droughts are no more common and are unlikely to become more common.

FLOODS

> *"Overall, there is no evidence that extreme weather events, or climate variability, has increased, in a global sense, through the 20th century, although data and analyses are poor and not comprehensive. On regional scales there is clear evidence of changes in some extremes and climate variability indicators. Some of these changes have been toward greater variability; some have been toward lower variability."*
> U.N. IPCC (1996a, p. 173)

The U.N. Intergovernmental Panel on Climate Change, or IPCC, is the international organization that is charged with proving that man-made CO_2 is the cause of global warming. That panel argues and propagandizes that global warming is occurring and is harmful. As the first several parts of this book have made clear, warming is generally beneficial, not harmful.

Dr. William Alexander was Emeritus Professor of the Department of Civil and Biosystems Engineering at the University of Pretoria in South Africa, and a former member of the United Nations IPCC Scientific and Technical Committee on Natural Disasters.

Alexander's research team has found wet and dry years correlate much better to movements in the solar system and activity on the sun itself rather than man-made climate change hypothesis. One of the greatest weaknesses in the arguments put forward by the man-made global warming lobby is that the projections of extreme weather events and the effects on agriculture are not firmly linked to projections of temperature change.

Alexander tested the alleged consequences of climate change on the environment, and he has found that the hypothesis just does not pan out.

"The whole climate change issue is about to fall apart — Heads will roll!" states Alexander. "Neither African floods nor droughts have increased in the period that climate change is alleged to have happened[135]."

Floods have been a part of our world since ancient times and are a common occurrence in many places on the planet. Germany offers a rich

---

[135] https://www.fcpp.org/pdf/FB051%20Will%20Alexander%20Climate%20Change%20and%20Africa260706%20with%20picsdraft%20edited%20-%20erin.pdf

history of flood documentation, with records dating back centuries. For instance, a German city named "Basserstand" experienced its worst flood in 1501, followed by notable floods in 1595 and 1954. This historical data suggests no discernible trend toward increasing floods in this region over time. The image below shows the worst of Basserstands' floods was in 1501. As is evidenced by the image, there has been no increase in flooding and nothing came close to the 1501 and 1595 floods until 1954, seventy years ago.

In the U.S., we have data in the image below from Great Falls National Park near Washington D.C. This "totem pole" looking waterline recorder shows no significant trend in flood occurrence since the 1960s. The worst floods in that area occurred in 1936, 1942, and 1972. The lack of a clear upward trend in flood occurrences challenges the notion that floods are becoming more frequent or severe.

A signpost at Great Falls with an image of flooding from the Potomac River.

The following image from ScienceAlert, an online science news publication, is titled, "An Incredible 45-Day Storm Turned California into a 300-Mile-Long-Sea." The article reveals that in 1861 for forty-three days, it rained almost nonstop in Central California. Rivers running down the Sierra Nevada Mountains turned into torrents that swept entire towns away.

In 1861, CO_2 was 287 ppm. Today we're closer to 420 ppm[136].

---

[136] https://www.sealevel.info/co2.html

Chapter Eighteen: How Earth's Weather is Made 235

The Great Flood of 1862, which affected the western U.S., remains unmatched in its scale and impact. According to meteorologist Jan Null:

> "...heavy rains to fall statewide just shy of the proverbial '40 days and 40 nights.' Los Angeles got 35 inches of rain, the foothill town of Sonora was pummeled with 102 inches and the Sacramento River crested at 24 feet above normal. It was a extreme weather disaster, long before any impact of the CO2 being blamed today for far milder weather[137]."

We should not have a concern of droughts or floods. They are a part of the Earth's natural weather system and beyond our control or influence;

---

[137] https://californiahistoricalsociety.org/blog/the-great-flood-of-1862/

other than through building dams, dikes, irrigation, and other projects that can have a real impact. The good news is that we are safer now than ever before; floods happened in the past and will happen in the future, but they're statistically less of a threat than in past centuries, according to clear data.

FIRES, FIRE PROPAGANDA, AND STATISTICS

> *"Research funding for environmental research in Australia, in my case mercury and wildfires, is almost impossible unless it is part of yet more greenhouse data gathering. There is also an atmosphere of intimidation if one expresses dissenting views or evidence. It is as if one is doing one's colleagues a great disservice in dissenting and perhaps derailing the gravy train[138]."*
>
> Dr. David Packham

Dr. David Packham is a distinguished figure in the field of climate research and meteorology. With a background as a former principal research scientist at Australia's CSIRO and as a senior research fellow in a climate group at Monash University, he brings a wealth of experience and expertise to his work. Additionally, Packham has served as an officer in the Australian Bureau of Meteorology, further solidifying his credentials in the field. His contributions to climate science and meteorological research have been invaluable in advancing our understanding of these complex phenomena.

In a related context, an American scientist Patrick T. Brown, recently admitted to "leaving out the full truth" to get a climate change wildfire study published in a journal.

Brown, climate team co-director at the nonprofit Breakthrough Institute in Berkeley and a visiting research professor at San Jose State University, said his Aug. 30 paper[139] in the prestigious British journal "Nature" is scientifically sound and "advances our understanding of climate change's role in day-to-

---

[138] https://www.climatedepot.com/2010/12/08/special-report-more-than-1000-international-scientists-dissent-over-manmade-global-warming-claims-challenge-un-ipcc-gore-2/

[139] https://www.nature.com/articles/s41586-023-06444-3

day wildfire behavior..." However, Brown also wrote that the study didn't look at poor forest management and other factors that are just as, if not more, important to fire behavior because "I knew that it would detract from the clean narrative centered on the negative impact of climate change and thus decrease the odds that the paper would pass muster with 'Nature's' editors and reviewers[140].

Brown also said that this focus on climate "misinforms the public" and "makes practical solutions more difficult to achieve[141]."

Brown is a prominent climate scientist with a diverse academic and research background. He currently serves as the Co-Director of the Climate and Energy Team at The Breakthrough Institute and holds an adjunct faculty position at the Energy Policy and Climate Program at Johns Hopkins University. Brown earned his Ph.D. in Earth and Climate Sciences from Duke University, complemented by a master's degree in meteorology and climate science from San Jose State University and a Bachelor's degree in Atmospheric and Oceanic Sciences from the University of Wisconsin – Madison.

Contrary to popular belief, the trend for forest fires and wildfires is decreasing, not increasing as has been the claim made by propaganda and leftist politicians.

Researchers Detect a Global Drop in Fires

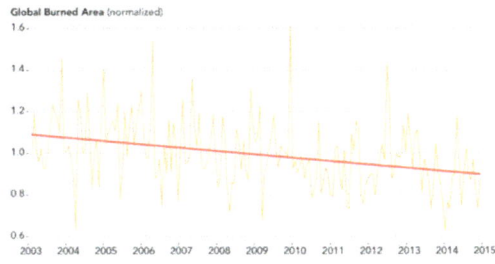

One of the most interesting things researchers have discovered since MODIS began collecting measurements, noted Randerson, is a decrease in the total number of square kilometers burned each year. Between 2003 and 2019, that number has dropped by roughly 25 percent.

---

[140] https://phys.org/news/2023-09-scientist-left-full-truth-climate.html
[141] https://phys.org/news/2023-09-scientist-left-full-truth-climate.html

The primary factor influencing wildfires is the presence of fire fuel on the ground and its dryness. Many regions worldwide practice controlled burning of forests and clearing of dead, dry brush and wood. Historically, indigenous peoples in America used controlled burns to manage forest ecosystems effectively. Fires are also a means of clearing land for agriculture or grazing in the poor subsistence farming areas of the world.

California's wildfire policy, which allows forests to remain untreated and lacks controlled burns, often results in devastating fires. In contrast, other states like Texas, South Carolina, Wisconsin, and Idaho, with similar forested areas, experience fewer wildfires due to better forest management practices.

In India, agricultural practices like burning crop stubble contribute to significant air pollution. Encouraging the adoption of gas-powered agricultural equipment would reduce both CO_2 emissions and harmful pollutants, as well as improve crop harvests and quality of life for everyone involved, even people in cities miles away who breathe the smoke-filled air.

The Biden Administration is actively manipulating wildfire data for propaganda purposes, aided by complicit media and government agencies. They exclude historical records and omit crucial information to fit their climate change narrative.

In the image below by NASA RECOVER/Keith Weber, you'll see that there has been a steady increase in the number of fires in the western U.S. over the past 60 years. In fact, the majority of western fires — 61 percent — have occurred since 2000 (shown in the graph below).

However, NASA is omitting the full data set to make it appear that fires in the western U.S. have increased. In fact, forest fires in the U.S. used to be far higher.

Over the past six decades, there has been a steady increase in the number of fires in the western U.S. In fact, the majority of western fires—61 percent—have occurred since 2000 (shown in the graph below).

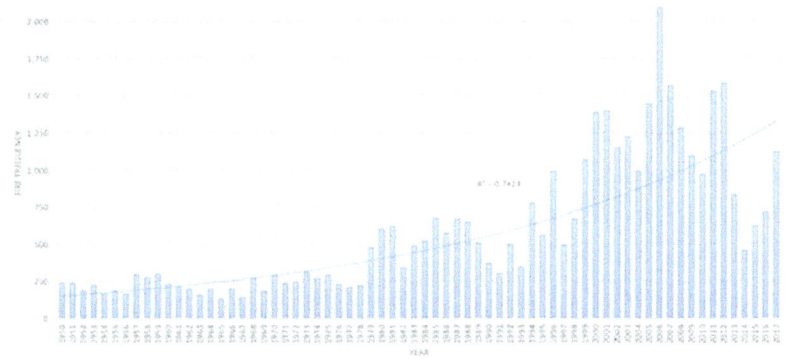

Source: NASA RECOVER/Keith Weber

In the following graph by the EPA, you'll see clearly that the actual amount of U.S. forest area burned between 1926–2017 has decreased measurably. This data was removed after Biden took office.

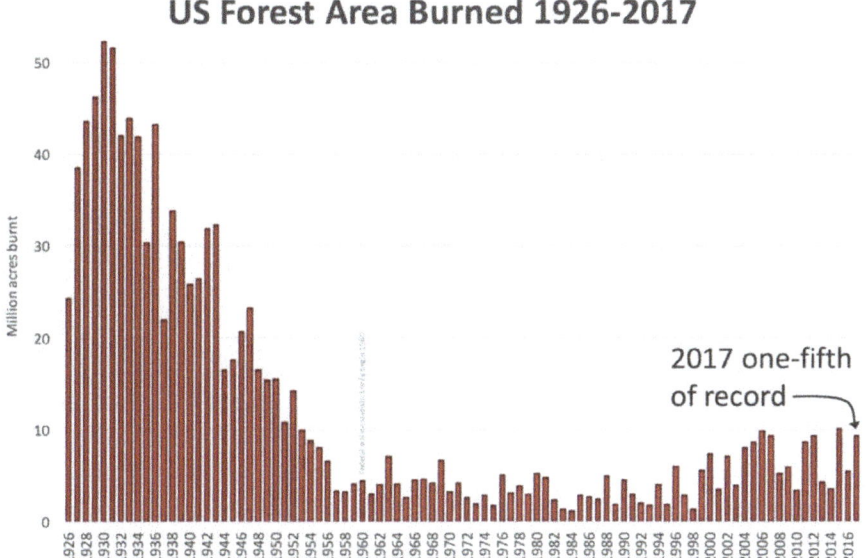

The Wayback Machine and climate heretics like me keep and make copies of this and so many other graphs, charts, images, resolutions, documents, and interviews on the internet because big tech scrubs them, corporations change data, and universities or government institutions have data that doesn't fit their narrative manipulated or even erased. Once we bring up paper clippings and old graphs, people like me doing research are ridiculed. Even though we're just presenting the truth.

As of January 29, 2021, the National Interagency Fire Center (NIFC) showed burn acreage all the way back to 1926. You'll see that burn acreage was much higher prior to 1960. And yes, we do have good data from 100 years ago on fires in the United States[142].

The Washington Post cherry-picks data to fit their story as well. In this graph from January 29, 2021 entitled "National Interagency Fire Center Burn Acreage," you'll see that *The Washington Post* only showed data from 1960 to 2016, hiding the exponentially larger numbers of fire burn acreage from before 1960. In essence, the data shown is a sliver that fools the eye into seeing an increase that appears like a small bump in the landscape. Take a step back and you'll see a 14,000-foot mountain that you can't see from the thin slice that was published.

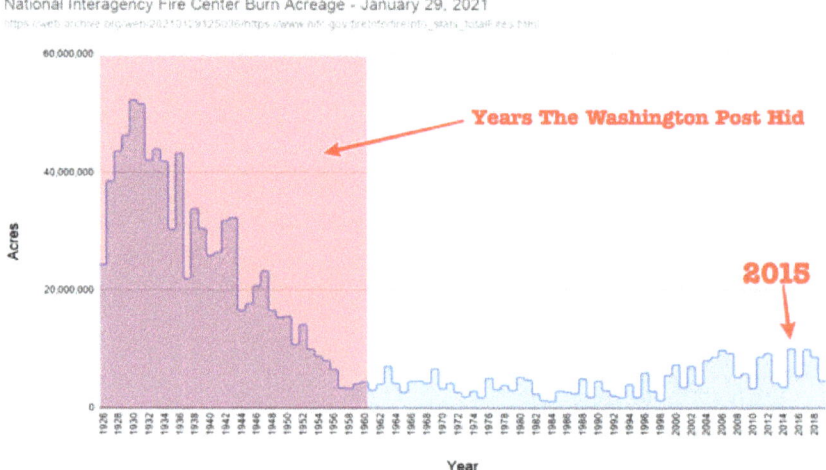

[142] https://web.archive.org/web/20210129125036/https://www.nifc.gov/fireInfo/fireInfo_stats_totalFires.html

Chapter Eighteen: How Earth's Weather is Made

In short, NIFC started their graph in 1983 in order to hide the fact that we used to have a lot more fires. What's more, they got rid of the explanation that we didn't have good data. Also, some time after President Joe Biden took office, the data prior to 1983 was erased. Biden has a full court propaganda press at all the federal agencies. The link now redirects to a boldface lie on the "National Interagency Fire Center" that states on the page, "Prior to 1983, the federal wildland fire agencies did not track official wildfire data using current reporting processes. As a result, there is no official data prior to 1983 posted on this site."

We do, in fact, have good data prior to 1983 or even prior to 1960 in the United States. "They" also claim we put a man on the moon in the 1960s... but we didn't have good data on fires? It wasn't the Dark Ages.

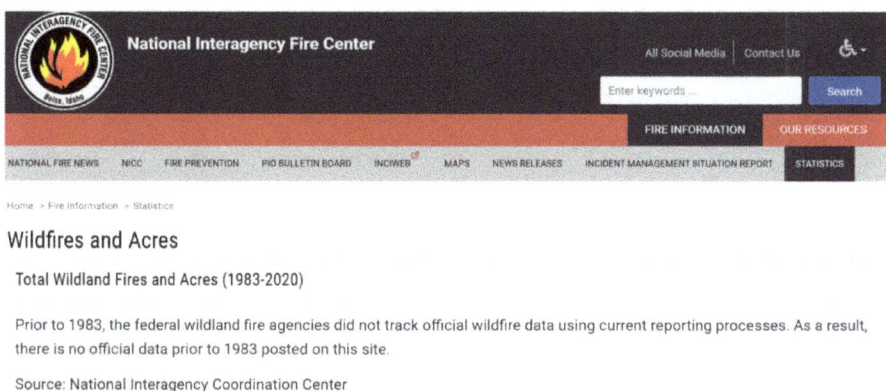

We know that in 1871, a series of devastating fires swept through various parts of the U.S., leaving widespread destruction and loss of life in their wake. Among the most notorious incidents were the Great Chicago Fire and the Peshtigo fire in Wisconsin. Legend has it that the Chicago fire started when a cow belonging to Mrs. O'Leary kicked over a kerosene lamp, igniting a blaze that quickly spread through the city. The fire razed large portions of Chicago, destroying thousands of buildings and leaving an estimated 300 people dead.

Large parts of Michigan were also burned by wildfires. These fires, fueled by dry conditions and strong winds, contributed to the overall destruction and chaos experienced across the region. The newspapers of the time described the Midwest as a tinderbox before the fires took place; it was hot and dry back then.

The most effective approach to mitigating wildfires involves proactive forest management, such as controlled burns or manual fuel removal. These practices are essential for reducing the risk of catastrophic fires and safeguarding communities and ecosystems.

The following image by my organization at "Truth in Energy and Climate" shows the difference between federally-controlled land and state-controlled land. The federally-controlled land is neglected with mostly dead trees. It's an extreme fire risk as well as a social, economic, and environmental liability. The right side of the image depicts state-controlled land which is actively managed with mostly live trees presenting a low fire risk and a social, economic, and environmental asset.

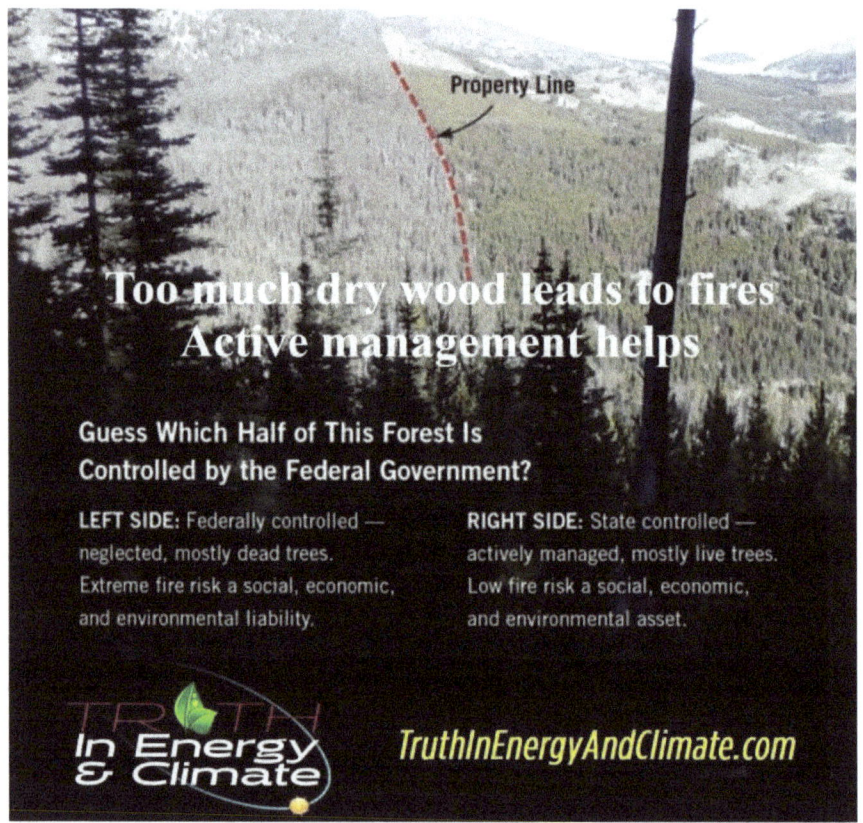

www.truthinenergyandclimate.com

Forest fires are a natural part of ecosystems; and yet, eight out of ten are caused by humans. Addressing wildfires requires a balanced approach that includes effective forest management strategies and realistic assessments of climate influences. Blaming climate is an easy out for government officials that are not doing their job.

Let's spread the truth, like wildfire!

Part Five:
THE WORLD IS SAFER THAN EVER

CHAPTER NINETEEN:
ORGANISMS INCLUDING HUMANS ARE THRIVING

"Global warming reduces more deaths than it causes, saving possibly 100,000 lives each year..."
Bjorn Lomborg

Each year, more than 100,000 people die from cold temperatures in the U.S. and 13,000 in Canada — which is more than 40 cold deaths for every one death from heat. On a worldwide basis, cold deaths vastly outnumber heat deaths. This is not just true for colder countries like Canada but also warmer countries like the U.S., Spain, and Brazil. Even in India, cold deaths outnumber heat deaths by seven to one. Globally, every year about 300,000 deaths are caused by heat, whereas almost 1.7 million people die of cold[143].

Dr. Bjorn Lomborg is a prominent figure in the field of economics and environmental policy. Leading the Copenhagen Consensus think tank, he collaborates with esteemed economists and Nobel Laureates to identify and advocate for effective solutions to global issues like disease, hunger, education, and climate change.

Lomborg has been widely recognized for his contributions...he was named one of *TIME Magazine*'s 100 Most Influential People in the World. He is also a bestselling author, with notable works including *False Alarm: How Climate Change Panic Costs Us Trillions, Hurts the Poor, and Fails to Fix the Planet*; *The Skeptical Environmentalist*; and *Cool It*.

In an article in The Financial Post Lomborg states:

---
[143] https://www.businesslive.co.za/bd/opinion/2021-08-01-bjorn-lomborg-a-one-sided-climate-focus-leads-us-badly-astray/

A landmark study in Lancet shows that across every region, climate change has brought a greater reduction in cold deaths over the past few decades than it has caused additional heat deaths. On average, it has avoided upwards of twice as many deaths, resulting in perhaps 200,000 fewer cold deaths each year. Apart from not mentioning that higher temperatures reduce deaths from cold, reporting that emphasizes the need to cut CO_2 usually pushes some of the least effective ways to help future victims of heat and cold. At best, climate policy will slow the increase in heat deaths only slightly. But we already know much more effective and simpler ways to help[144].

The Lancet Study cooks the "cold and death" graph. In the following image, the left illustrates deaths from cold; they use 50 per 100,000 deaths. On the right with the same scale, it is 10 deaths per 100,000 people, so the numbers don't look so lopsided. They still are lopsided, however, with far more dying of cold.

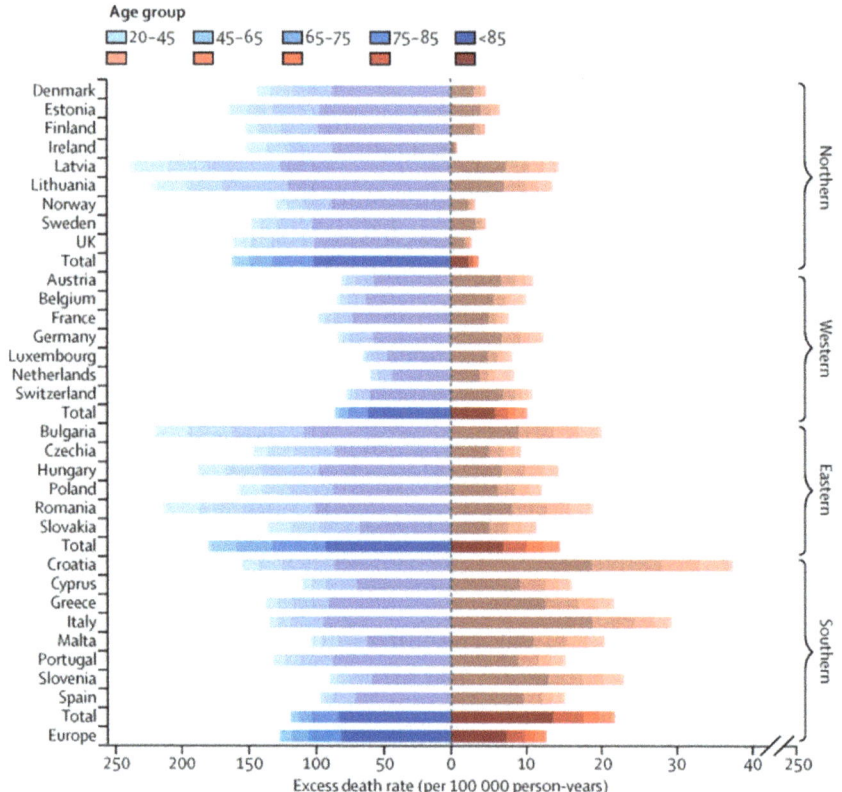

Source https://www.thelancet.com/journals/lanplh/article/PIIS2542-5196(23)00023-2/fulltext

---

[144] https://financialpost.com/opinion/bjorn-lomborg-climate-change-and-deaths-from-extreme-heat-and-cold

Chapter Nineteen: Organisms Including Humans Are Thriving 249

Lomborg's think tank the Copenhagen Consensus Center shows an accurate, not misleading graph, in a public letter. It shows heat deaths are stable and far lower than cold deaths. And cold deaths are down more than 26% over the last 30 years. Instead of 6 times more cold deaths 30 years ago, now it's just 4 times more. This is because more people can afford heating, from more affordable coal, oil, and natural gas.

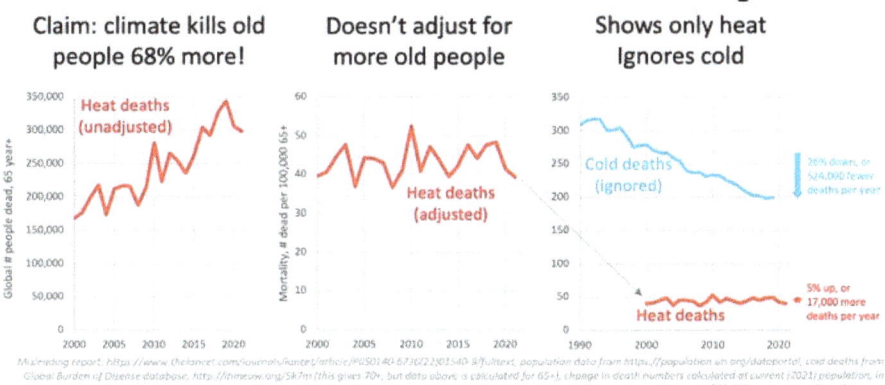

Source https://twitter.com/BjornLomborg/status/1586697882702323713

Despite the global population reaching eight billion–a record high–we are experiencing fewer deaths from extreme weather events than ever before. This is largely due to advancements in technology that allow us to forecast storms and hurricanes with greater accuracy, giving us more time to prepare and protect ourselves.

Additionally, our knowledge of construction techniques has improved, enabling us to build stronger and more resilient infrastructure that can withstand severe weather conditions.

The progress we have made in implementing disaster preparedness measures, such as concrete bunkers in vulnerable areas, and controlling stormwater runoff has significantly reduced the loss of life during extreme weather events. Investments in infrastructure, such as dikes and dams for flood control and electricity generation, have further contributed to our ability to survive natural disasters.

It's important to recognize that affordable, abundant, and reliable energy plays a crucial role in improving peoples' lives and enabling progress. Rather than focusing on limiting essential resources like CO_2, which is necessary for plant growth and sustenance of life on Earth, we should prioritize investments in disaster preparedness and resilience-building efforts.

Contrary to popular belief, trends in tornadoes, hurricanes, floods, and droughts show no significant increase according to scientific assessments, such as those by the Intergovernmental Panel on Climate Change (IPCC). Moreover, data from the UN Office for Disaster Risk Reduction (UNDRR) indicates a decline in the number of disasters over the past two decades, suggesting that there is no imminent climate crisis. In the graph below you'll see that the number of disasters by subgroup (e.g., geophysical, hydrological, meteorological, and climatological) per year has remained fairly steady since 2000, even dropping slightly.

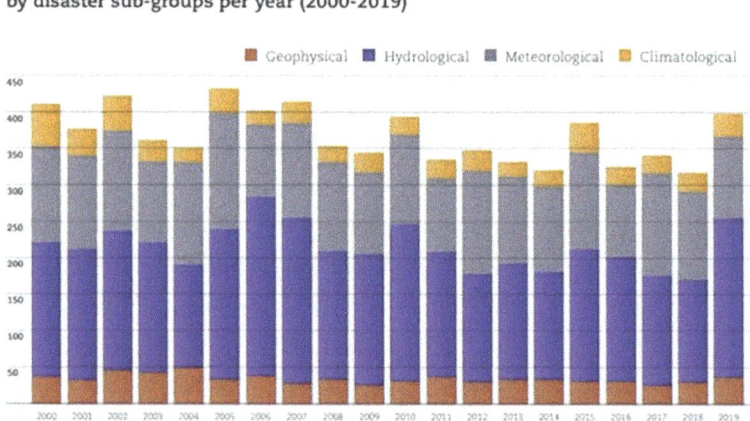

Graph Courtesy of Our World In Data[145]

Despite the global population quadrupling over the past 120 years, fatalities resulting from weather-related events have significantly decreased. The graph below by "Our World In Data" (sourced originally from EMDAT OFDA/CRED International Disaster Database, Universite Catholique de

---

[145] https://ourworldindata.org/natural-disasters

Louvain in Brussels, Belgium), shows clearly that over the past 100 years global annual death rates from natural disasters have simply plummeted. Deaths by impact, landslide, mass movement (dry), and drought aren't even calculable by this method since before the 1900s.

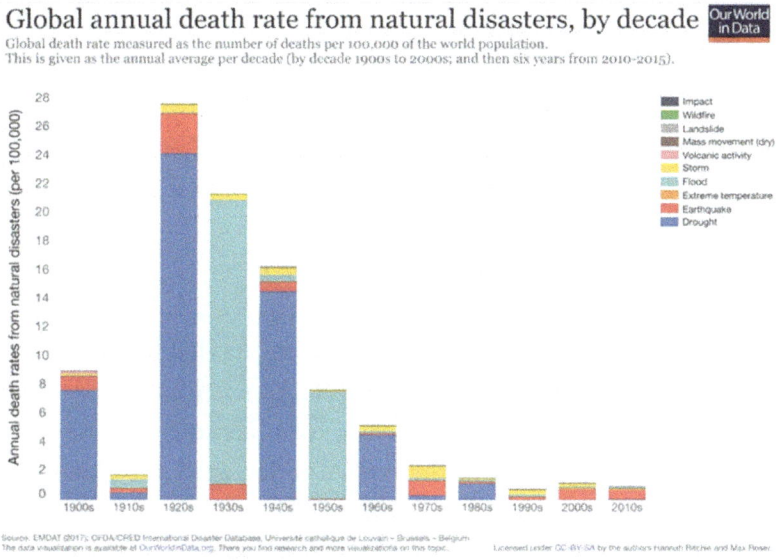

While Lomborg acknowledges the role of human-generated CO_2 in climate change, he diverges from the narrative that it poses an imminent catastrophe. Instead, he emphasizes the importance of prioritizing cost-effective solutions to address pressing global challenges.

Lomborg argues that excessive focus on reducing CO_2 emissions diverts resources from more impactful endeavors aimed at improving human well-being.

Given these trends and advancements, there is currently no evidence to suggest that the frequency or severity of extreme weather events will suddenly worsen in the foreseeable future. We are safer now than ever from weather events.

The world is becoming a safer place for people, largely because of warming. We are better off in a warmer world with plentiful, affordable fossil fuel energy available to more people. They fuel the machines and we make products from them to protect and enhance our lives. We use this fuel to build homes and other structures that can keep us safe from weather events and natural disasters.

LESS DISEASE WITH BETTER LIVING CONDITIONS

> *"Elimination and eradication are the ultimate goals of public health. The only question is whether these goals are to be achieved in the present or some future generation."*
> Walter R. Dowdle, Ph.D

Walter R. Dowdle, Ph.D, a virologist by training, has held several significant roles in public health. He served as the CDC Assistant Director for Science and Director of the newly formed Center for Infectious Diseases. In 1986, he transitioned to the Office of the Assistant Secretary for Health (HHS) in Washington, where he established the Office of Coordinator for AIDS Activities, Public Health Service (PHS). He also served as Deputy Director of the CDC and Deputy Administrator of the Agency for Toxic Substances and Disease Registry (ATSDR).

Dowdle states: "Elimination and eradication are the ultimate goals of public health. The only question is whether these goals are to be achieved in the present or some future generation[146]."

Louis Pasteur, a famous French chemist, pharmacist, and microbiologist acclaimed for his groundbreaking discoveries in vaccination, microbial fermentation, and pasteurization pioneered research in chemistry that led to insights into disease and prevention. His work is credited for advancements in hygiene, public health and modern medicine. Pasteur states, "It is within the power of man to eradicate infection from the earth[147]." My publisher shared with me that these were some of the last words also spoken by the late Dr. Vladimir Zelenko, whose groundbreaking discoveries in the

---

[146] https://www.ncbi.nlm.nih.gov/pmc/articles/PMC2305684/
[147] Dubos, René Jules, and Jean Dubos. The white plague: tuberculosis, man, and society. p228. Rutgers University Press, 1987.

"Zelenko Protocol" reveal methods for treating every conceivable single-strand RNA virus, including Covid-19[148] with drugs such as Ivermectin and Hydroxychloroquine, as well as several over-the-counter alternatives.

The decline in infectious diseases is a testament to advancements in healthcare, sanitation, and disease management practices rather than solely attributable to changes in climate. While some argue that a warmer world could potentially expand the habitat of disease-carrying organisms, such claims often overlook the broader ecological dynamics at play.

For instance, as growing seasons extend further north, it not only facilitates the spread of certain diseases but also fosters the proliferation of beneficial plants and animals into new regions. This aspect is overlooked in climate propaganda, which focuses only on negative outcomes.

Progress in combating diseases like malaria and river blindness underscores the effectiveness of targeted interventions and medical treatments. Through initiatives such as population-based treatment with drugs like Ivermectin, significant strides have been made in eliminating these diseases in various regions worldwide.

Countries like Colombia, Ecuador, Mexico, and Guatemala have successfully eradicated Onchocerciasis, a parasitic disease known as River Blindness, through dedicated elimination efforts, demonstrating the importance of coordinated healthcare strategies.

The mischaracterization of medicines like Ivermectin as "unsafe" or "ineffective" highlights propaganda at its worst; despite its proven efficacy and safety profile, misinformation surrounding its use has persisted, underscoring the need for evidence-based approaches to both healthcare communication and climate communication.

Chloroquine has indeed been a widely used medication for preventing and treating malaria in areas where it is endemic. It has also been employed in the management of conditions such as amebiasis, rheumatoid arthritis, and lupus erythematosus (under appropriate medical supervision).

While there have been discussions about the safety and efficacy of chloroquine for various indications, particularly in the context of its off-label use for COVID-19, it remains an important tool in the prevention

---

[148] https://rumble.com/v3r9a11-the-zelenko-protocol-with-frank-zelenko-thl-ep-22-tues-oct-24-12pm-pt-live.html

and treatment of malaria, especially in regions where the disease is prevalent. This medicine is safe and has been prescribed to millions who live in and travel to malaria-infested areas of the world.

Overall, chloroquine has demonstrated a favorable safety profile when used appropriately, and it continues to play a valuable role in global efforts to combat malaria and other related diseases. Yet, we were told it was dangerous. Why? This is similar to the misinformation and disinformation that we see spreading about climate change.

In summary, while climate can influence disease dynamics to some extent, the primary drivers of disease prevalence and control are multifaceted and involve factors such as healthcare infrastructure, sanitation practices, and access to medical interventions. Effective disease management requires a comprehensive understanding of these factors and targeted interventions to address them.

The world's people would be far better off investing in more healthcare infrastructure, sanitation practices, and access to medical interventions rather than wasting money on trying to stop the climate from changing, because as I've mentioned before, the climate has always changed and always will.

MOST ANIMALS AND INSECTS ARE DOING FINE THERE IS NO MASS EXTINCTION

The narrative told to us by climate alarmists of a looming sixth mass extinction is unsubstantiated and driven by propaganda, rather than evidence.

Historically, extinctions have been primarily caused by factors such as habitat loss, introduction of non-native species, and disruption of delicate ecosystem balances. For instance, the introduction of non-native species like rats, cats, and snakes to remote islands led to a significant number of species extinctions during the 1500s and 1600s.

Today, human mobility continues to facilitate the movement of plants, animals, insects, and bacteria to new regions, disrupting native species and altering environments. While some introductions have had positive outcomes, such as the stocking of non-native fish in lakes for sports fishing, others have resulted in ecological imbalances, like the proliferation of pythons in the Everglades.

Unsustainable practices, not climate change, are the real villains in creating vulnerability in several of the Earth's ecosystems. Of the forests in the Democratic Republic of Congo, the World Resources Institute states, "A major contributor to this loss is the unsustainable exploitation of forests to meet growing charcoal demand. Charcoal is produced by cutting and burning timber. Logs are stacked in traditional, low-efficiency kilns where the high heat turns them into charcoal. An expanding population with a growing need for food, energy and economic development has led to the rapid growth of the charcoal industry[149]."

However, there are success stories demonstrating the resilience of species when they are protected and their habitats are restored. For example, sea turtles in Seychelles and whale populations in Australia have rebounded from the brink of extinction due to conservation efforts.

Data from the Red List, which identifies species at risk of extinction, shows a declining trend in extinctions over time. This underscores the effectiveness of habitat protection and species conservation in facilitating population recovery.[150]

These recent headlines show some of the many success stories:

1. "How we discovered that sea turtles in Seychelles have recovered from the brink"
   March 17, 2022[151]

2. "Whales' rebound in Australia a success story: Duke biologist"
   April 8, 2022[152]

---

[149] https://www.wri.org/insights/how-charcoal-industry-threatens-drcs-forests
[150] https://www.iucnredlist.org/
[151] https://theconversation.com/how-we-discovered-that-sea-turtles-in-seychelles-have-recovered-from-the-brink-179041?utm_source=Nature+Briefing&utm_campaign=c764af37f4-briefing-dy-20220318&utm_medium=email&utm_term=0_c9dfd39373-c764af37f4-46184474
[152] https://coastalreview.org/2022/04/whales-rebound-in-australia-a-success-story-duke-biologist/

This article illustrates that when we protect species and their habitats, they usually repopulate. As habitat is restored, animal numbers often expand.

3. "Wildlife Is Recovering in Europe After Decades of Conservation"
September 27, 2022
Bison, elk, beavers, and many bird species have become more abundant, according to a report by Rewilding Europe[153]. The article states, "There's more than 50 species coming back to Europe in the past 50 years, but by looking at these we can see what worked," Schepers said. "It shows that, if you take measures, wild animals can recover."

4. "Endangered tigers making a 'remarkable' comeback"
July 29, 2020[154]
From this BBC article: "In 2010, there were as few as 3,200 wild tigers. But now five countries — India, China, Nepal, Russia, and Bhutan — have given hope for the future.

Conservation efforts and habitat preservation are crucial for maintaining biodiversity and ensuring the well-being of both humans and wildlife. As societies become more prosperous, they often develop a greater appreciation for their environment and have more resources to invest in conservation efforts. They move from hand to mouth to being able to afford to care about their environment and do something positive about it.

Transitioning away from traditional, environmentally harmful fuels such as dung, wood, charcoal, and crop waste for heating and cooking is essential for reducing pressure on natural habitats. The availability of abundant, cleaner energy sources like coal, kerosene, propane, natural gas, and electricity enables this transition and contributes to the restoration of natural areas.

Improving access to cleaner cooking and heating fuels not only benefits the environment but also enhances the quality of life, particularly for women

---

[153] https://www.bloomberg.com/news/articles/2022-09-27/wildlife-is-recovering-in-europe-after-decades-of-conservation?sref=2rMwa6IQ
[154] https://www.bbc.com/news/newsbeat-53581028

who bear the burden of collecting fuel and are exposed to harmful smoke during cooking.

The Dominican Republic has subsidized propane use for cooking and heating, because of the country's economic reliance on fossil fuels. This has helped them with cleaner fuel, and has also led to the protection of forests and natural habitats. The image below courtesy of the National Forestry Corps illustrates the difference between the Dominican Republic and Haiti; it shows that the DR side is verdant and green, while the Haitian side is dry and almost void of any green hues.

National Forestry Corps ...
btlonline.org

The example of Hispaniola illustrates the stark contrast between the Dominican Republic and Haiti in terms of environmental sustainability and economic prosperity. While the Dominican Republic has maintained lush forests and a vibrant ecosystem through the use of fossil fuels and enlightened conservation policies, Haiti has faced severe deforestation and economic challenges due to reliance on wood charcoal for energy.

The disparity between the two nations underscores the importance of implementing sound energy policies, fostering conservation efforts, and promoting sustainable development practices to improve the lives of people and protect natural habitats and wildlife.

In Haiti, deforestation is a pressing environmental issue, with significant consequences for biodiversity and ecosystem health. The reliance on wood charcoal for energy production has led to extensive forest loss over the years. From 1923 to 2006, Haiti experienced a dramatic decline in forest coverage, with less than 2% of the land remaining forested by 2006.

The Dominican Republic demonstrates a more favorable situation in terms of forest preservation: approximately 41% of its land remains forested, according to the U.N. Food and Agriculture Organization. These forests play a crucial role in sequestering carbon and supporting diverse ecosystems, housing numerous species of amphibians, birds, mammals, and reptiles. The contrasting outcomes in forest conservation between Haiti and the Dominican Republic demonstrates the importance of effective environmental policies, sustainable land management practices, and access to cleaner energy sources in order to mitigate deforestation pressures and promote ecological resilience.

This story could be replicated all over the world. But the Net Zero or "no CO_2 released by man" initiatives will not allow this to happen. Our resources would be much better spent on moving all countries from wood, charcoal, dung, and crop waste to fossil fuels sooner, rather than attempting to change the future climate by limiting the life-giving trace gas CO_2.

Overall, while challenges remain, the assertion of an imminent mass extinction event lacks factual support. Mass extinctions are a fear-mongering tactic, rather than an accurate reflection of biodiversity trends.

THERE'S LESS POVERTY IN THE WORLD NOW THAN EVER BEFORE

"Fossil Future starts with a premise few would dispute: Fossil fuels helped create the modern world. Access to low-cost energy has been responsible for lifting millions of people out of abject poverty, making inhospitable environments livable, and supplying the goods and services that underlie modern standards of living. In recent human history, energy has largely meant fossil fuels. As a result, we're just not ready to turn off the taps

tomorrow—fossil fuels are part and parcel of the world we've made."
Alex Epstein

Alex Epstein is an American author, energy theorist, and industrial policy expert. He is known for his work advocating for fossil fuels and defending the use of hydrocarbons. He wrote the book *The Moral Case for Fossil Fuels*, which argues that fossil fuels have improved the quality of life for billions of people and are morally praiseworthy. Epstein is also the founder and president of the Center for Industrial Progress, a think tank that promotes industrial progress and energy freedom. He frequently writes and speaks on topics related to energy, environmentalism, and industrial policy.

Epstein makes his moral claim, "The goal for energy policymakers... should be to enable 'human flourishing'—human beings' ability to live long, healthy, fulfilling lives[155]."

As Epstein sees it, access to cheap energy is what makes this possible. I found his case compelling well before ever seeing him speak or reading his books, and that admiration is now consecrated.

Affordable, reliable, abundant energy makes for less poverty and more prosperity. Mix in sovereignty and free speech, the rule of law, private property rights, limited bureaucracy, low levels of graft, and CO_2 for more plant growth and you have a recipe that ensures mankind's future.

Metrics across the board indicate longer, better lives for people, both in absolute numbers due to the global population having reached 8 billion, and as a percentage of the total population.

There's something downright ominous about people who promote depopulation or claim "overpopulation." They're essentially promoting murder, at least implying that they promote death. In Paul Ehrlich's infamous 1968 book, *The Population Bomb*, he predicted widespread starvation and food shortages. None of his catastrophic forecasts materialized, but he continues to peddle alarmism to this day. Even recent appearances on mainstream platforms like "60 Minutes" propagate his erroneous claims about a supposed 6th mass extinction event, which as I've proven here, is far from reality.

---

[155] https://foreignpolicy.com/2022/05/28/fossil-fuels-climate-change-energy-epstein-review/

Contrary to these fearmongering narratives, the world is experiencing real, tangible, positive progress. Poverty rates have plummeted, with only around 10% of the global population living in poverty today compared to 85% in 1900 and 75% in 1950. This remarkable improvement should be a cause for celebration!

If climate discourse was not turned upside-down, humanity's achievements would be celebrated – or at least acknowledged – and leveraged for further progress. The following graph entitled "Share of People in and Out of Poverty (1820-2011) illustrates that the percentage of people not living in absolute poverty has plummeted, with the most significant decrease happening after about 1950[156].

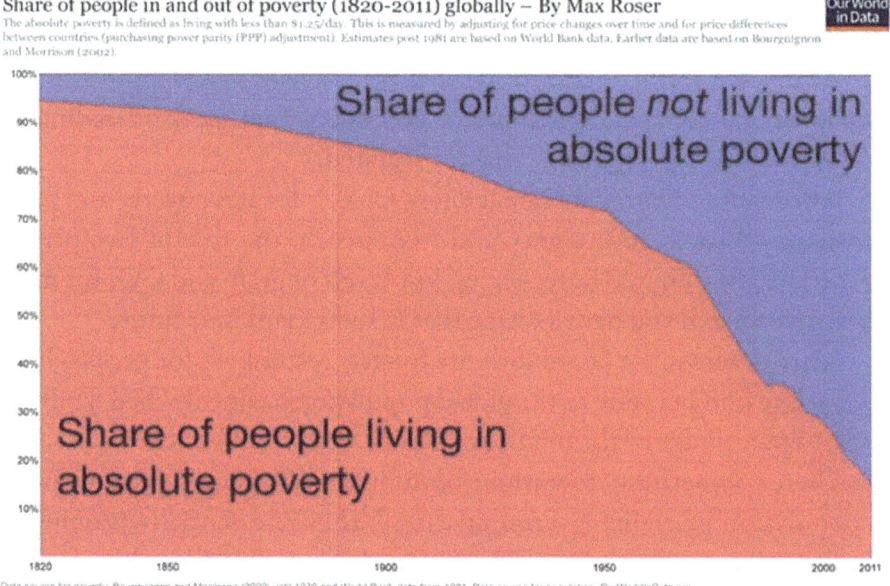

Despite the good news about poverty, the climate reality denying crowd is doing the bidding of the global and political elite, and in light of the evidence, it's hard not to label them as downright Luciferian. These population-hating elite advocate for policies that will reverse mankind's

---

[156] https://ourworldindata.org/extreme-poverty-in-brief

progress through climate energy and food policies. Soon, they will be advocating anti-freedom policies to control what we eat and consume, to limit CO_2 to supposedly save us from a climate that doesn't need saving. Don't believe me? A cursory glance at www.worldeconomicforum.com will tell you all you need to know; I encourage you to check it out, and please share your findings widely.

Misguided climate policies in countries like Sri Lanka and South Africa have set back progress in their countries. South Africa has let their coal electricity plants degrade so badly, that many have been closed. They have not built any more to replace the closed ones, either. Coal in South Africa is abundant and low cost.

Harvesting would generate economic opportunities for poor people in that country. South Africa has attempted to build more industrial wind and solar capacity. However, because it takes three times more wind and five times more solar to replace a coal plant, progress has been slow. Because wind and solar are part-time and undependable, you still need full-time generation to supply electricity all the time.

The price South Africa is paying for their climate-driven energy policy amounts to nationwide blackouts across the country, often for 8 to 12 hours daily. Can you imagine not having electricity for half the day, every day? These blackouts are hurting their people and their economy. Is this part of the global elite's plan for the West next?

Access to affordable, reliable energy is crucial for human prosperity. Fossil fuels have played a pivotal role in lifting millions out of poverty, transforming inhospitable environments into habitable ones, and driving modern standards of living. Energy, predominantly derived from fossil fuels in recent history, fuels progress and innovation. Without fossil fuels we will lose the prosperity we have gained over the last 50 years.

Machines make human progress happen. Only with energy can those machines really multiply the ability of people to do more, create more, and make their lives better. And just as important, they make their kids' and grandkids' lives better.

The data overwhelmingly supports the notion that the world is getting better, not worse. Instead of fixating on doomsday scenarios, we should leverage our successes and continue building a prosperous future for all.

It's imperative to engage with younger generations, instilling optimism and equipping them with real facts about the world's progress. Affordable energy, coupled with freedom and increased CO_2 for plant growth, promises a bright future where humans can continue to flourish and build.

I urge you to talk with the under 30 crowd; help them understand the facts and regain optimism for the future. We need them to participate in helping make the Earth an even better place for all people, animals, and wildlife.

ALL IS WELL WITH THE CLIMATE - CLIMATE POLICY IS THE REAL DANGER

The prevailing narrative around climate change paints a grim picture of the future unless we take extreme action. The truth is far more optimistic. Contrary to popular belief, the Earth's climate has been gradually warming since the end of the Little Ice Age ended about 1850. This warming trend is not alarming and started long before CO_2 could have been the cause. Increased CO_2 levels and warmth are *good* things.

As we all learned in Part Two, CO_2, demonized by the climate truth deniers, is necessary for plant growth. More CO_2 grows better plants. Worldwide crop harvests have seen significant increases, largely attributed to rising CO_2 levels, leading to a greener world and enhanced food production. In fact, natural gas-derived fertilizers, powered by fossil fuels, have played a critical role in feeding billions of people worldwide.

We should embrace more CO_2 in the atmosphere and the huge benefits it brings.

Despite this progress, certain policies pushed by politicians and the global entities UN and WEF, threaten to undermine these advancements. Proposals such as carbon rationing and limitations on farming and fertilizer use are not only unnecessary, they are also harmful to global prosperity.

Carbon rationing, in particular, will lead to widespread downgrading of our lifestyles unnecessarily. Rationing will increase inequality, disproportionately affecting the middle class while allowing the wealthy elite to continue their lavish lifestyles unchecked. Similarly, restrictions on farming and fertilizer use would drive up food prices and create shortages, directly contradicting the evidence that increased CO_2 levels have bolstered crop yields.

It's crucial to recognize these misguided efforts for what they are: attempts to control the masses and consolidate power in the hands of a select few. By spreading awareness and advocating for sensible policies that prioritize freedom, prosperity and real fact-based environmentalism, we can push back against this dark agenda.

We must reject fearmongering and instead focus on solutions that uplift all people and foster a brighter future for generations to come. We must engage and work toward meaningful change based on facts, not fear. We must end the climate driven anti-human agenda of the UN, WEF, and global elites. We can do this by spreading this good news; sharing this book, these studies, this insight you've gleaned from reading or listening to this collection of facts.

Affordable, abundant, reliable energy and our food supply are in the line of fire from the power-hungry globalists and their climate cult followers. We need to protect ourselves, our children, and their children. There are enough resources for all to prosper. We don't need to endanger our entire way of life based on the climate change lie.

Part Six:
ENERGY

CHAPTER TWENTY:
IT USED TO BE DARK – IT STILL IS IN MANY PLACES

Energy has been a driving force behind the progress of human civilization from the earliest sources of energy like candles and whale oil lamps, to modern marvels like electricity and space travel; energy has enabled us to illuminate the darkness, power our homes and businesses, and propel us to incredible heights of technological achievement.

Coincidentally, the whales are recovering with conservation efforts; they were originally saved by kerosene becoming widely available for reading light, replacing whale oil. I'm personally thrilled we no longer use their fat for lamps.

The transition from traditional sources of energy like tallow candles to more efficient and abundant forms such as paraffin and kerosene marked significant advancements in human history. These advancements not only brought light to the darkness but also paved the way for broader accessibility and affordability of energy, improving the quality of life for millions around the world.

The following graph from Our World Data shows how coal, displaced traditional biomass for energy. Then oil added more energy, joined by natural gas, and only in the last decades have wind and solar added small percentages. In 1800 the world used 5653 TWh of energy, in 2022 it 179,000 TWh. We use nearly 32 times more energy now than we did in 1800. We have about eight times more people. We use four times more energy per person…some of us many times more than that. And there are still billions of people living with the same or not much more energy than was used in 1800.

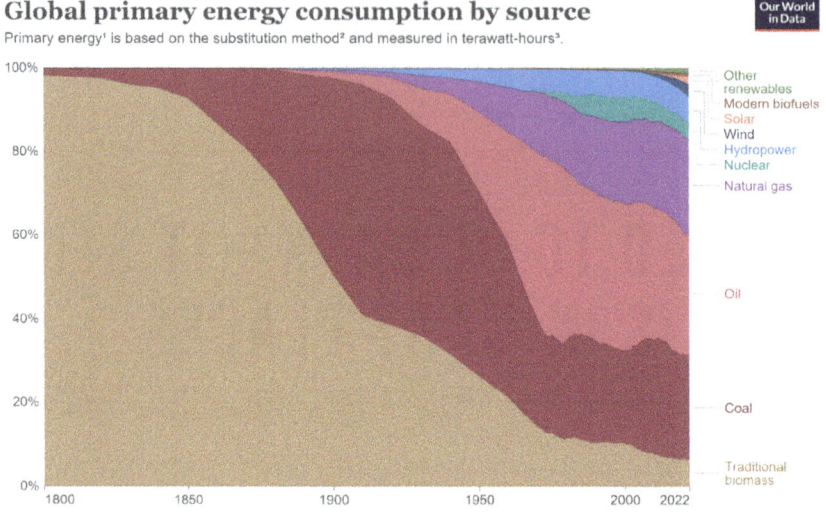

https://ourworldindata.org/energy-mix

Today, our dependence on energy is more pronounced than ever. From powering our homes and vehicles to fueling the technology and infrastructure that underpin modern society, energy is at the heart of nearly everything we do. The expansion of electrification, the advent of space exploration, and the proliferation of digital technologies all show the indispensable role of energy in shaping the world as we know it.

Yet, despite the tremendous strides we've made, there are still billions of people worldwide who lack access to reliable and affordable energy sources. Addressing this disparity and ensuring that everyone has access to clean, efficient energy should be a top global priority. By empowering individuals and communities with the means to harness energy for their needs, we can improve living standards, drive economic development, and foster greater resilience in the face of challenges.

Moreover, as we continue to pursue advancements in areas like artificial intelligence, renewable energy, and sustainable development, it's essential to recognize the foundational role that fossil fuels have played, and continue to play, in supporting our civilization. While there is a growing push toward wind and solar energy sources, we must understand the reality that fossil fuels remain an integral part of our energy landscape, providing the necessary foundation for progress and prosperity.

In essence, energy is the lifeblood of modern civilization, fueling our innovations, sustaining our communities, and enabling us to thrive in an ever-changing world.

Vaclav Smil is a renowned interdisciplinary researcher whose work spans a wide range of critical topics, including energy, environmental sustainability, population dynamics, food production, innovation history, risk analysis, and public policy. With an impressive body of work that includes 47 books and over 500 papers published as of June 2023, Smil has established himself as a leading authority in these fields.

As a Distinguished Professor Emeritus at the University of Manitoba and a Fellow of the Royal Society of Canada, Smil's deep understanding of complex issues and his ability to distill dense information into accessible insights make his work invaluable to scholars, policymakers, and anyone interested in understanding the inner workings of our world.

In his recent book, *How the World Really Works,* Smil continues his tradition of offering profound and illuminating perspectives on key global issues. With his characteristic depth of analysis and incisive observations, Smil provides readers with a nuanced understanding of the interconnected systems that shape our lives. Whether delving into the intricacies of energy production, the dynamics of population growth, or the challenges of sustainable development, Smil's work offers invaluable insights that challenge conventional wisdom and provoke critical thinking.

For those who want to deepen their understanding of pressing global challenges and explore innovative solutions, Smil's work serves as an indispensable resource. You will be drawn to his rigorous research methodology, his rich historical context, and his forward-thinking policy recommendations in *How the World Really Works*. Smil states:

> *"Four materials rank highest on the scale of necessity, forming what I have called the four pillars of modern civilization: cement, steel, plastics, and ammonia are needed in larger quantities than are other essential inputs. The world now produces annually about 4.5 billion tons of cement, 1.8 billion tons of steel, nearly 400 million tons of plastics, and 180 million tons of ammonia. But it is ammonia that deserves the top position as our most important material: its synthesis is the basis of all nitrogen fertilizers, and without their applications it would be impossible to feed, at current levels, nearly half of today's nearly 8 billion people[157]."*

In a *New York Times* interview, Smil stated: "*Now, according to COP26 (2021) we should reduce our carbon dioxide emissions by 45 percent by 2030 as compared with 2010 levels. This is undoable because there's just eight years left, and emissions are still rising. People don't appreciate the magnitude of the task and are setting up artificial deadlines which are unrealistic.*[158]"

Energy is the lifeblood of modern civilization, powering everything from transportation and industry to agriculture and healthcare. Without sufficient energy, our way of life would grind to a halt, plunging us into darkness and depriving us of the comforts and conveniences we take for granted. It's no wonder these fears have been preyed upon!

Fossil fuels, including coal, oil, and natural gas, have long been the backbone of our energy supply, providing the vast majority (83%) of the world's energy needs. Their abundance, reliability, and energy density have propelled our civilization forward, enabling unprecedented economic growth, technological innovation, and improvements in living standards.

The following graph from Our World Data shows us visually where the world's energy, in 2022 comes from. It shows that in terms of percent of the world's primary energy, oil was 29, coal 25, natural gas or methane was 22, hydropower was 6.3, traditional biomass (wood, dung, and crop waste) was 6.2, wind was 3, and solar was just 2 percent. The climate cultists say that in 25 years (2050) we should end oil, coal, and natural gas and somehow replace them with wind and solar. That is a long way to go in a short amount of time. It is expensive, nonsensical fantasy thinking. And China and India, with 2.8 billion people between them have different plans.

---

[157] https://time.com/6175734/reliance-on-fossil-fuels/
[158] https://www.nytimes.com/interactive/2022/04/25/magazine/vaclav-smil-interview.html

Chapter Twenty: It Used to be Dark – It Still Is in Many Places

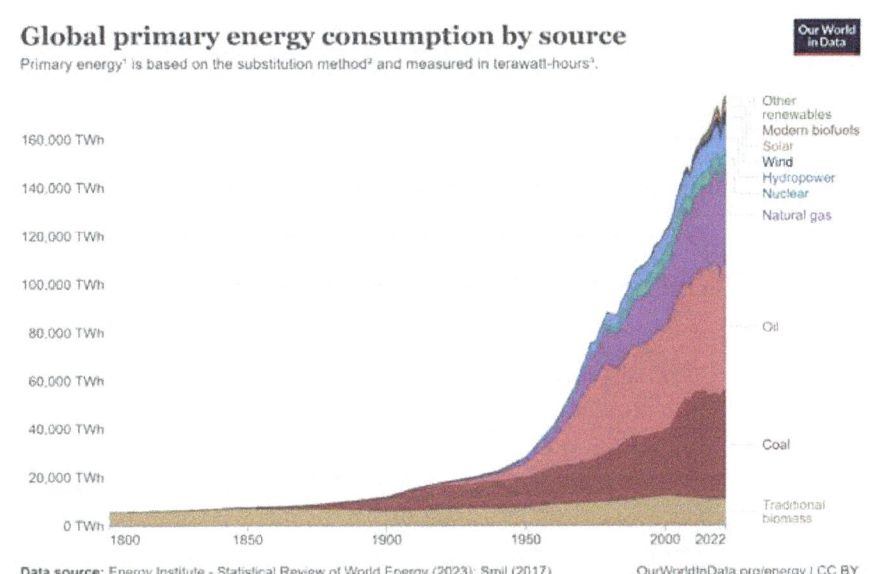

https://ourworldindata.org/global-energy-200-years

Yet, despite their undeniable importance, fossil fuels have come under attack in recent years, vilified by climate alarmists who claim they are driving catastrophic climate change. Climate policies aimed at phasing out fossil fuels in favor of intermittent energy sources like wind and solar are being promoted as the solution to solve the false climate crisis.

Transitioning away from fossil fuels is neither practical nor feasible. Renewable energy sources, while valuable in certain contexts, are unable to provide the reliable, on-demand power required to sustain modern society. Their intermittent nature, low energy density, and the amount of land they require, make them ill-suited to meet the growing energy demands of a growing global population.

Moreover, the economic consequences of phasing out fossil fuels would be devastating, leading to higher energy costs, reduced economic growth, and widespread poverty. Already, we are seeing the impacts of climate policies on energy prices, with consumers paying more at the pump and in their electricity bills.

Furthermore, the push to eliminate fossil fuel-derived fertilizers threatens global food security, jeopardizing the ability of billions of people to access affordable, nutritious food. The example of Sri Lanka, where food production plummeted following the banning of synthetic fertilizers, serves as a stark warning of the consequences of such misguided climate policies.

In the face of these challenges, it is imperative that we reject the alarmism and propaganda peddled by climate activists and instead focus on pragmatic solutions that prioritize human well-being and economic prosperity. Rather than sacrificing our energy security on the altar of climate ideology, we must embrace a balanced approach that leverages the full range of energy sources available to us.

We should stop subsidizing wind and solar energy. We cannot continue to add to our $34 trillion-dollar national debt and $1 trillion annual interest payments. At some point, it will bankrupt our nation. Every dollar we spend on wind, solar, transmission wires for them, EVs, and green subsidies is adding to our debt. Worse, we are purchasing most of these green things from China, with money we don't have. And we are making our electric grids unreliable and far more expensive.

ABUNDANT AFFORDABLE RELIABLE ENERGY MATTERS

Abundant, affordable, and reliable energy is required for prosperity. It dramatically improves the quality of life for people around the world, making the entire community safer and healthier as a result. As energy consumption has increased over the past century, so have all the indicators of human well-being such as life expectancy, income levels, and access to basic necessities.

The following image entitled "The World As 100 People Over the Last Two Centuries" was originally published by Professor Max Roser[159] on the website OurWorldInData.org in 2016[160]. The graph shows us that gains in living standards have elevated global living conditions over the past 200 years. Free markets (see Chapter One), democratic limited government capitalism, and advances in energy technology are all to thank for these wins

---

[159] https://www.maxroser.com/
[160] https://ourworldindata.org/a-history-of-global-living-conditions-in-5-charts/

for humanity. Here are some things we learn from this data as summarized by Professor Mark J. Perry[161]:

1. In 1820, 90% of the world population lived in extreme poverty vs. only 10% today.
2. In 1820, 83% of the world population had not attained any education vs. 14% today.
3. In 1820, 88% of the world population was illiterate vs. only 15% today.
4. In 1820, 99% of the world population was not living in a democracy vs. 44% today.
5. In 1820, none of the world population was vaccinated against diphtheria, whooping cough, and tetanus vs. 86% today.
6. In 1820, 43% of the world's children died before age five vs. only 4% today.

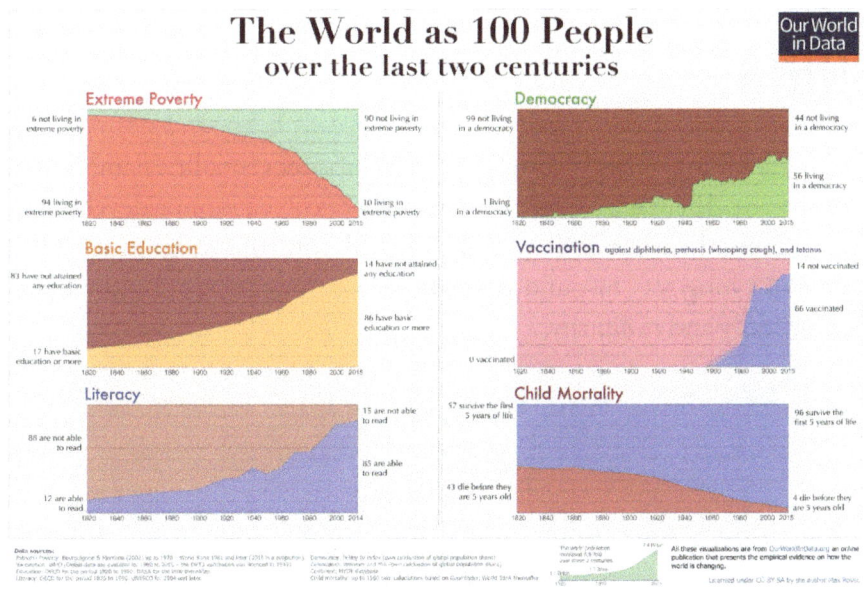

The correlation between energy technology advancements and prosperity is undeniable, as demonstrated by the graphs depicting the simultaneous

---

[161] https://www.aei.org/carpe-diem/the-world-as-100-people-over-the-last-200-years-a-period-of-the-largest-gains-in-global-living-standards-ever/

growth in population, individual wealth, and energy consumption. This relationship illustrates the importance of ensuring that energy remains accessible and affordable for all, particularly where raising living standards are still needed, which is most of the non-Western world.

Blanket opposition to fossil fuels or unrealistic mandates for renewable energy sources undermine the very foundations of global prosperity and freedom. Efforts to transition to wind and solar are unnecessary, expensive, and will not be sufficient enough to replace the energy from oil, natural gas, and coal any time soon. The attempt will turn out to be a "green" nightmare.

By prioritizing access to affordable, reliable energy, we can work toward a world where everyone has the opportunity to thrive and prosper. Keeping this out of the reach of billions of people holds them back and is bad for Earth. India, China, and other countries are working to increase their energy availability, reliability, and affordability for their people. They understand that in order to have prosperity, they must increase the availability of energy.

CHINA

Looking at China's energy landscape and its broader geopolitical implications is important. There is a complex interplay between economic development, environmental sustainability, and political control. Communist China has come a long way fast and is set up to continue its rapid growth, both economically and militarily.

Tiananmen Square, photo courtesy of @axe65 for Adobe

China's rapid economic growth over the past few decades has been fueled by a massive expansion in energy consumption, particularly from coal. Despite efforts to diversify its energy mix, coal remains the dominant source of energy in China, making up over half of its total energy. This heavy reliance on coal has significant environmental consequences, contributing to air pollution. Many people can be seen walking around big cities in China wearing masks outdoors to prevent inhaling the toxic smog pollution from coal burning plants.

China has made strides in adding wind and solar power, but these sources still represent a small fraction of their overall energy mix.

The chart from Our World Data shows that China gets 55 percent of all its energy from coal, 18 percent from oil, 8.5 percent from natural gas, 7.7 percent from hydro-power, 4.5 percent from wind, and just 2.5 percent from solar. So much for the energy transition in China.

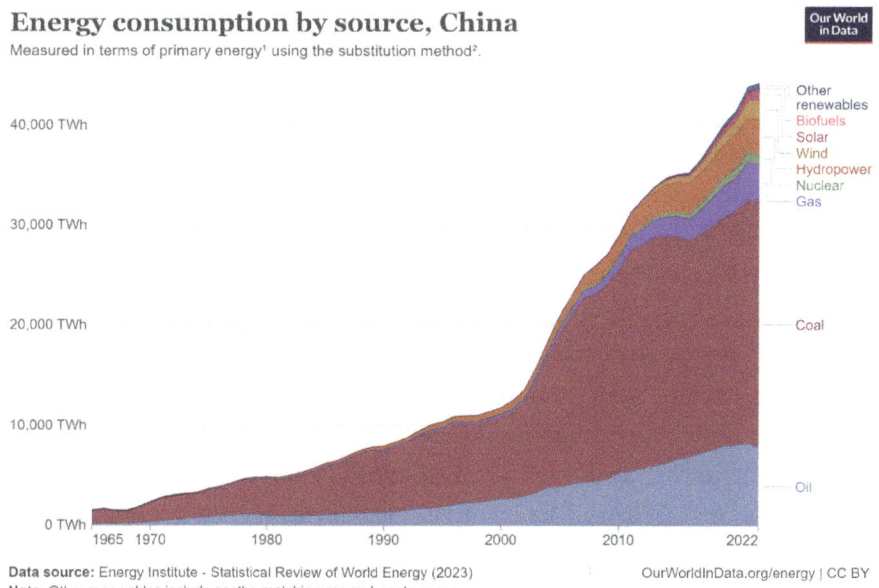

https://ourworldindata.org/energy/country/china

The rest of the worlds' wind turbines, solar panels, batteries, electric vehicles, and the rare metals needed to make them are either built or processed in Communist China. While these items are the cause of much virtue signaling in the West, many of them are processed with forced or slave labor, dirty coal plants (without clean coal technology), and very little environmental protection, if any.

The Chinese government's tight control over information and dissent has enabled it to prioritize economic growth and political stability over environmental concerns and human rights. The social credit system, which monitors and penalizes citizens based on their behavior, exemplifies the extent to which the government exercises control over its populace.

They have facial recognition cameras everywhere and control their people through their phones. They keep track of citizens' behavior by holding a "social credit score" over their heads, which is weaponized in order to force citizens to behave according to CCP rules. Criticize the communist party in a public place and you are likely to have your social credit score downgraded. If your social score gets too low, you cannot buy anything, travel, or even use the Internet. You must be a good communist, or else.

Any critical thinking or freedom thoughts can only be expressed in private in China, and even then, you are in danger of being found out and punished. They also have a spy network of others to rat you out, similar to Eastern Europe under the Soviet Union regime. It also resembles Mark 13:12 in the Bible, where the New American Standard Version reads: "Brother will betray brother to death, and a father his child; and children will rise up against parents and have them put to death[162]."

China's aggressive foreign policy and expansionist agenda, particularly in the South China Sea and Taiwan, pose significant risks to regional stability and international security. The country's growing influence in global markets coupled with its strategic investments in critical infrastructure abroad raise valid concerns about its long-term intentions. China is a bully to other nations as well. Many corporations are being attacked by Chinese malware which can destroy infrastructure, take companies offline, and hinder profits.

---

[162] https://bible.knowing-jesus.com/topics/Disobedient-To-Parents

This is part of a larger national security threat in which Chinese companies seek to cripple competition through cyber-attacks.

Meanwhile, China is expanding its military power while stockpiling coal and food supplies. According to Farrelly & Mitchell, "a few countries control the biggest share of food stockpiles. According to the USDA, China holds half of the world's wheat reserves and 70% of its corn... The U.S. has 6% and 12% of global wheat and corn reserves, respectively[163]."

What are they preparing for?

There have been recent instances of severe weather and energy shortages in China. However, they clearly weren't stockpiling coal to help their people; they kept large portions of their stockpiles in reserve, while people suffered from weather-caused adversity. They're saving it for something else... and your guess might be as good as mine as to what... I have some ominous ideas.

China has a long history of experiencing intermittent bad weather: droughts, floods, too hot, or too cold. When there are droughts, they are not able to produce electricity from their hydroelectric dams. This causes shortages, and they must substitute with other fuel sources, such as coal.

From an energy perspective, China being the world's largest consumer and producer of coal, makes any global efforts to stop fossil fuels a moot point. We know that ending fossil fuels is foolish and unnecessary. China certainly knows this, and is proceeding accordingly.

China is building more than 655 coal plants right now. Each of those plants lasts 50–75 years. As of the time of this writing, they are on a veritable coal electricity plant building spree, and I doubt China will be closing these brand-new coal plants any time soon. China emits more CO_2 and pollution than all other 28 industrial nations combined, including the United States.

Efforts to promote transparency, accountability, and respect for human rights in China is a must. Yet, the left and other cronies of the cult of climate change don't seem to care about this. Could it be that there is a kinship among China's leaders and the Wests' leftists who lean toward socialist and communist ideologies? Remember, history has shown us that communists stick together, even if it means millions starve to death, lose their freedoms, or "have" to be killed for the good of communism.

---

[163] https://farrellymitchell.com/our-thinking/latest-agribusiness-blog/chinas-food-stockpiling-the-rise-in-food-prices

Our American socialists wrongly say that real socialism hasn't been implemented anywhere, ever, and that they would do a better job than Cuba, Venezuela, the former Soviet Union, or Pol Pot in Cambodia. Socialism has never worked and never will. Don't confuse it with the great welfare states of Scandinavia which are now fraying because of the influx of foreigners who don't share their native culture's sense of needing to contribute to society in exchange for generous state and taxpayer provided benefits.

Ultimately, the trajectory of China's energy transition will have far-reaching impacts for the world. Chinas' energy habits continue to shape the future of global energy markets, geopolitical dynamics, and environmental sustainability (or the lack thereof). From a national security perspective, this should concern us all. As a key player in the global economy and the world's largest emitter of pollution and greenhouse gases, China's actions will profoundly impact our future.

Communist China is a force to be reckoned with; a real and credible threat to our way of life, national security, and world peace. In a decade or two, they will be the largest economy in the world. Unfortunately, they have a history of using all that is at their disposal to get their way, not being fair, and even bullying other nations to get their way. Our energy policies that cause reliance on China are not in our best interest, nor are they in the best interest of the Earth.

INDIA

India now has more people than China, each with close to 1.4 billion. Life expectancy in India has doubled since 1950, from 35 years to 70 years[164,165]. India's per capita income went from $368 and 1990 to $2100 in 2019[166]. They want the strong growth that China has, which will require lots more energy. Therefore, India is committed to bettering the lives of its people and increasing the strength of its nation with more energy[167].

---

[164] https://www.worldometers.info/world-population/india-population/
[165] https://www.worldometers.info/world-population/china-population/
[166] https://data.worldbank.org/indicator/NY.GDP.PCAP.CD?locations=IN
[167] https://www.iea.org/reports/india-energy-outlook-2021/energy-in-india-today

Since 1990, India's electric usage has quadrupled from 330 kWh to 1,229[168] kWh per person. Electricity usage per person in the United States was 10,791[169] kWh. The average person in the U.S. uses nearly nine times more electricity than someone in India. India uses four times more energy of all sorts than they did in 1990. You can see the increases in the following graph.

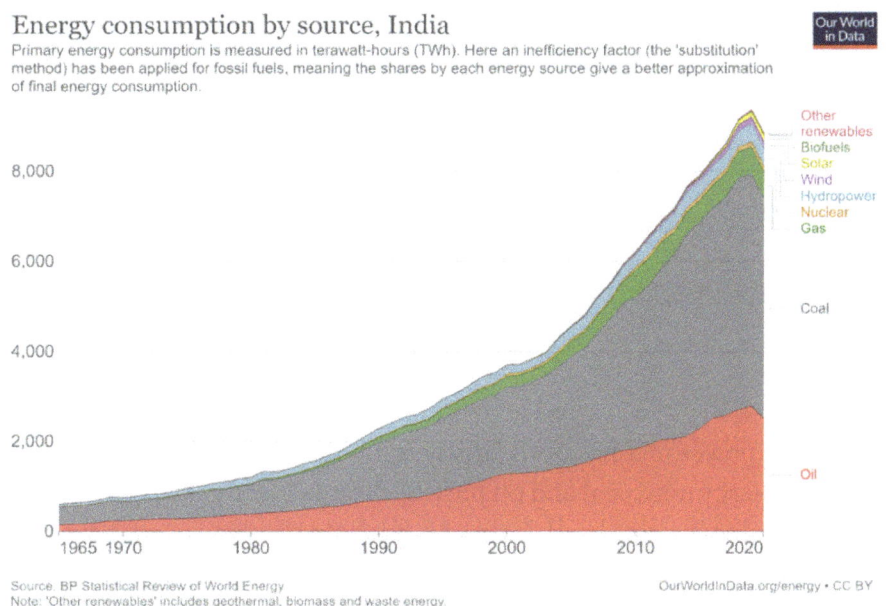

In most of the U.S., we typically enjoy 100% access to electricity, whereas up to 3% of households in India do not have electricity at all. 70% of households report their electricity was interrupted at least once last week, and large percentages have no electricity for an hour or more most days.

Like nearly every other country in the world, India gets the vast majority of their energy from oil, coal, and natural gas. We can see how much their energy use has grown from 1990 to 2022 in the graph above. Coal provides more than 50% of all their energy, oil is 30%, followed by natural gas at 7%. Wind and solar provide tiny amounts. The Our World in Data graph below shows this clearly.

---

[168] https://www.worlddata.info/asia/india/energy-consumption.php
[169] https://www.eia.gov/tools/faqs/faq.php?id=97&t=3

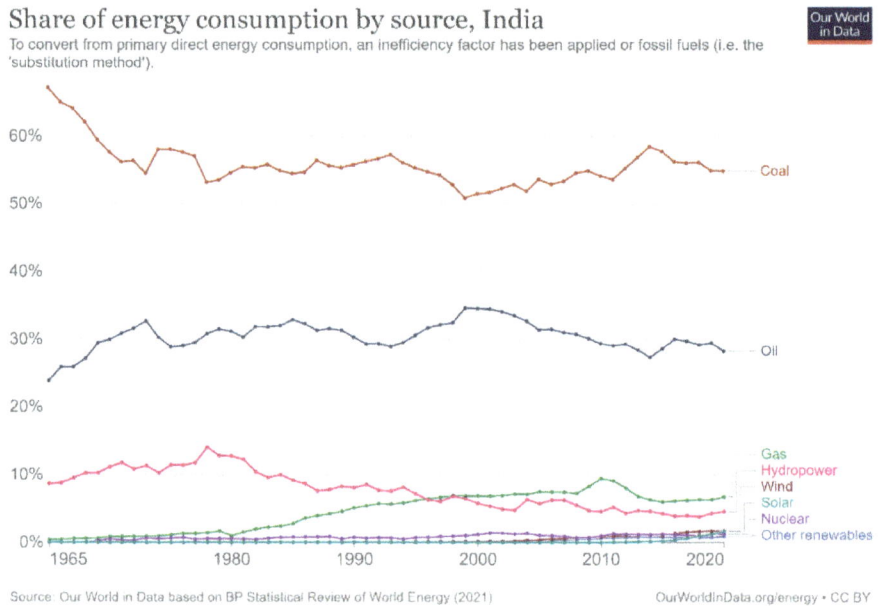

India's power mix has been largely static since 1990. While they use a lot more energy now, coal and oil provide 80%. Natural gas usage went up a little bit and then came back down so it has been the same percentage more or less. Wind power has actually declined as part of their energy mix as their economy has grown.

India's priority is to provide full-time electricity to 100% of their citizens[170]. They have taken the brakes off using coal to accomplish this because coal is reliable, low-cost, and abundant in India, yet they still import some, too.

Climate doomsters say that India needs to stop using coal, oil, and natural gas. India is doing the opposite. They want economic development, better lives for their people, and more power internationally. Coal is a major part of their strategy.

Does it make sense for the West to spend trillions on Net Zero when 2.8 billion people in these two nations are not? They are *ramping up* their fossil fuel energy and the climate cult is attempting to kill ours.

---

[170] https://www.forbes.com/sites/suparnadutt/2018/05/07/modi-announces-100-village-electrification-but-31-million-homes-are-still-in-the-dark/?sh=56b0cdbb63ba

COAL

The 8.5 billion tons of coal used worldwide every year provide as much energy as 41 billion barrels of oil. China uses more than half of those 8.5 billion tons. Coal usage is actually growing around the world. In most of the world, except for the West, it is the primary source of electricity generation because it is low cost and abundant in most countries.

Coal energy is used to make electricity and is a crucial component for various industrial processes such as steel and cement production. Its natural abundance, relatively low cost, and reliability make it a preferred choice in many regions, particularly in countries with extensive coal reserves.

The vast reserves of coal worldwide, estimated to be sufficient for hundreds to thousands of years of consumption, point to its importance as a long-term energy resource. As technology advances and extraction methods improve, access to coal deposits will become even more feasible and abundant.

Efforts to address environmental concerns associated with coal, such as air pollution, have led to advancements in clean coal technologies. There is a lot of hype concerning carbon capture and storage (CCS) and more efficient combustion processes. These technologies aim to suck up the CO_2 and put it underground "someplace."

Coal's versatility extends beyond energy generation, as it serves as a valuable feedstock for various industrial applications, including the production of carbon fiber, graphene, and building materials. This diversity of uses enhances the economic value of coal and underscores its importance in supporting industrial development and innovation.

However, it is essential to balance the benefits of coal with its environmental and social impacts, including detrimental air and water pollution. Sustainable management practices, regulatory frameworks, and investments in cleaner technologies are crucial for minimizing these impacts and ensuring the responsible use of coal resources.

Ultimately, coal will continue to play a significant role in the global energy landscape for the foreseeable future, providing a reliable and accessible source of energy for electricity generation and industrial processes.

Let's take a look at what Indiana states about their coal and energy development from the Indiana Office of Energy Development.

"Indiana coal mines are reclaimed through state and federal agencies such as DNR and Federal (DOI Office of Surface Mining) agencies. Post mining, most of this land is restored for wildlife habitat and water.

The U.S. has 22% of the world's proven reserves of coal, more than any other country.

Coal holds high value for other materials like carbon fiber, graphene, and building materials.

Coal's Operational Attributes

Electricity generated from coal is also cleaner now than ever before. Coal power plants have invested significant ratepayer dollars to drastically reduce emissions. The National Energy Technology Laboratory stated that a coal plant with pollution controls reduces nitrogen oxides by 83 percent, sulfur dioxide by 98 percent, and particulate matter by 99.8 percent compared to plants without controls.

Coal power plants, as any form of steam power generation, provide consistent and reliable delivery of electricity needs for Indiana's homes and businesses. Coal power plants that can produce electricity are considered baseload generation meaning they run all day, every day.

This is an important attribute which makes coal plants reliable, and reliability is critical, especially to support intermittent or variable energy resources.

Coal is abundant in Indiana and neighboring states thanks to the Illinois Basin. This makes coal a stable and cheaper source of energy since transportation costs are minimal.

> *Coal in Indiana is also used to support manufacturing, one of the state's strongest economic sectors. Coal is essential to the production of steel. The high-capacity factor and consistent heat rate in coal means it gives steel its strength, which is especially important for infrastructure like bridges, buildings, and automobiles[171]."*

As mentioned, China's extensive use of coal in its energy mix and industrial sector is well-documented, with hundreds of coal plants supporting its growing economy and meeting the demands of its large population. Similarly, other countries in Asia, such as India, Pakistan, and Indonesia, are also increasing their coal consumption to fuel economic growth and provide reliable energy access to their citizens.

The transition from traditional biomass fuels like wood, dung, and crop waste to modern energy sources like electricity, kerosene, natural gas, and propane can significantly improve air quality, public health, and living standards in developing countries. By electrifying rural areas and expanding access to clean cooking fuels, these nations can mitigate indoor air pollution and reduce reliance on environmentally harmful practices.

Moreover, the energy density of coal is substantial, providing a vast amount of energy equivalent to a significant volume of oil. This energy density combined with the ability to store coal on-site and access reserves in many regions globally ensures its continued relevance as an energy source for the foreseeable future.

---

[171] https://www.in.gov/oed/resources-and-information-center/about-indiana-resources/energy-fuels/fuel-facts-coal/

OIL

Oil remains a cornerstone of modern civilization, powering the vast majority of transportation systems and serving as a feedstock for numerous essential goods beyond just fuels. The versatility of oil-derived products spans from plastics and pharmaceuticals to fertilizers and synthetic fibers.

The U.S. uses 20 million barrels of oil a day, of the 101 million barrels used worldwide per day. We use about 37 billion barrels of oil a year. We use an incredible amount of this low-cost and reliable fuel source. The ongoing exploration and discovery of new oil and gas reserves will continue meeting global energy demand for many decades. The sheer scale of worldwide oil consumption ensures oil's enduring significance in the foreseeable future.

PRODUCTS MADE FROM OIL, COAL, AND NATURAL GAS

Coal, oil, and gas are some of the most important natural resources that we use daily. These fossil fuels are all hydrocarbons; they are compounds formed from the elements carbon and hydrogen. People around the globe rely on hydrocarbons for everyday products that are the foundation of modern society, from medical equipment to electronics, and they are derived from oil and other fossil fuels.

There are more than 6,000 different products made from oil, coal, and natural gas. We use them every day, from the toothbrushes we use in the morning to the polyester in our clothes, pillows, and sheets. Virtually all of our household items, like plastics and cleaning products, rely on one or more hydrocarbons. Most importantly, vital life-saving medical supplies and advanced technologies also rely on them. Modern society simply can't function without them. Hydrocarbons are deeply ingrained in our daily lives.

The following illustration shows that oil and other hydrocarbons give us medicine, cosmetics, plastics, synthetic rubber, cleaning products, and asphalt for our roads[172]. The OilFieldPulse.com graphic then lists scores of other products that we need hydrocarbons to make; things like TVs, cell phones, electric vehicles, and regular gas vehicles, and, frankly, pretty much everything you use is made from coal, oil, and natural gas.

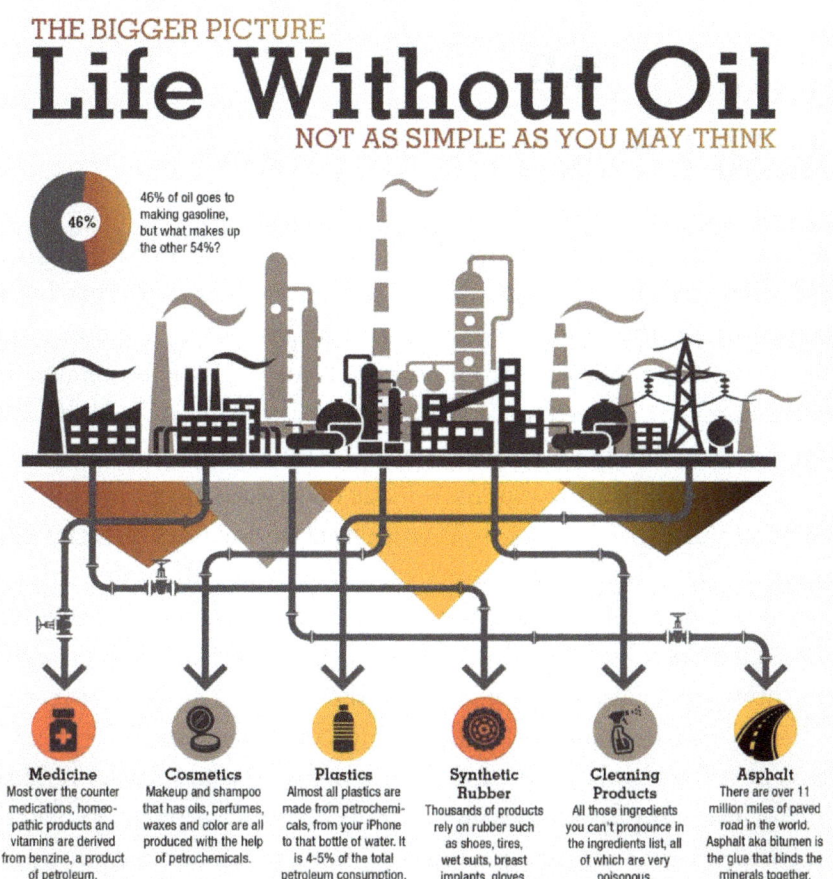

---

[172] https://i0.wp.com/wattsupwiththat.com/wp-content/uploads/2023/07/Life-Without-Oil-1688431851.9138.jpeg?ssl=1

Hydrocarbons brought about historical transformation with rapid widespread adoption. The transition from traditional energy sources to coal, then oil, and finally natural gas revolutionized human society, driving economic growth, technological advancement, and improvements in quality of life for everyone across the planet. Hydrocarbons aren't just for transportation.

NATURAL GAS (ALSO KNOWN AS METHANE)

The world uses a lot of natural gas. We have huge reserves of natural gas, and some believe that the Earth may be continuing to make natural gas, but that is controversial. The consensus believes natural gas is finite and that, at some point, several hundred years from now, we'll run out. My personal belief is that the Earth continues to make natural gas because the raw materials derive from the center of the Earth, where it is over 6000 degrees.

Titan, one of Saturn's moons, is made up of natural gas and oil. Space.com tells us that this one moon of Saturn holds more than the entire Earth's resources[173]. This leads me and many experts to believe that the Earth could continue to make oil as well as natural gas. This is also a very controversial opinion, and the consensus is that oil is finite, and that it is made from originally living things that have been buried under the Earth for millions of years.

The world uses about 140 trillion cubic feet of natural gas every year. The U.S. is now the number one exporter of natural gas, and Russia is number two. Because of the fracking revolution and directional drilling, we have been able to extract ever-increasing amounts of natural gas. Natural gas is moved around mainly by pipelines. It is turned into Liquid Natural Gas, LNG, for shipping, transport, and use in cars.

Natural gas provided about 40% of U.S. electricity needs in 2022[174], per the data collected by the U.S. Energy Information Administration, as shown below. Natural gas creates steam to churn giant turbines to make electricity. Natural gas has replaced a good amount of the coal generation for electricity.

---

[173] https://www.space.com/4968-titan-oil-earth.html
[174] https://www.eia.gov/energyexplained/electricity/electricity-in-the-us.php

Natural gas is a low-cost option because of fracking, and it is more easily turned up and down to match an electricity demand.

In addition, natural gas is made into thousands of different products, plastic being one of the most useful and ubiquitous. Natural gas makes nearly all hydrogen. It also makes fertilizers by taking nitrogen out of the air and turning it into usable fertilizer. This feeds half of the world, by increasing how much crop we can harvest from each acre, according to the website ourworldindata.org.

Like coal and oil, natural gas usage is also expanding around the world. It truly is a worldwide commodity; Asia intends to increase their use of all fossil fuels in order to expand their energy capabilities to improve citizens' lives.

ELECTRIC GRID FUNDAMENTALS

Electricity is a secondary source of energy that must be made from primary fuel sources. Over the past century, our electric grids have evolved into complex systems that we take for granted.

Electric grids lack the ability to store power, they must immediately meet demand with supply. Batteries are super expensive; grid scale batteries would cost trillions of dollars, and we simply don't have the manufacturing ability nor supply to make them within the decades-long time limit that the climate cult has deemed we will be allotted to switch to wind and solar. If we don't have enough electricity, then we'll have blackouts or curtailments. Curtailments are events during which electricity is selectively shut off to manage shortages.

Understanding the limitations of wind and solar power is crucial. While touted as renewable sources, they cannot provide the grid with enough electricity on demand, and they must be replaced every 20 to 30 years. That will be cripplingly expensive. The idea of replacing traditional power generation with renewables overlooks the unreliable problem of wind and solar. There isn't a good answer as to what will provide electricity on dark, cold, windless nights.

Wind turbines, towering at 70 stories with massive blades, typically generate electricity around 30% of the time, when wind conditions permit.

Solar panels fare even worse, producing electricity only 20-25% of the time and considerably less during winter months or peak demand periods. This means that they produce far less than their nameplate capacity.

When 200 Mw of solar are built, they only make 40-50 Mw of electricity. A 1,000 Mw of wind produces 300–330 Mw of electricity. This means we will need 300,000 more wind towers and 10,000 square miles of solar panels (or more), and that doesn't solve the part-time problem. We will also need thousands of miles of expensive transmission wires.

Operating a hybrid grid of natural gas, coal, and nuclear power alongside intermittent renewables is more expensive than steady, full-time power plant operations. Similar to how cars get better mileage on highways than in stop-and-go city driving, traditional power plants emit fewer pollutants and use less fuel when operated consistently.

The U.S. electricity mix shows that nearly 40% was sourced from natural gas, nearly 20% from coal, 18.2% from nuclear power, and around 13.5% from wind and solar combined in 2022.

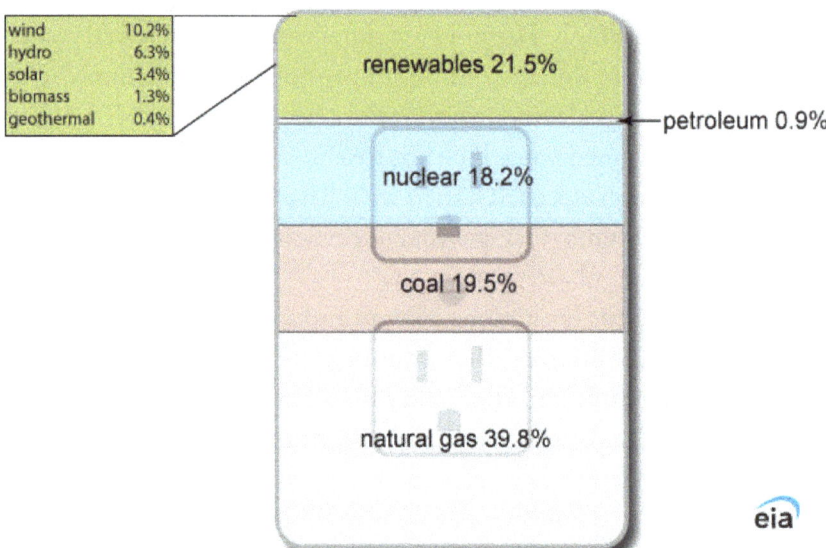

Data source: U.S. Energy Information Administration, *Electric Power Monthly*, February 2023, preliminary data
Note: Includes generation from power plants with at least 1,000 kilowatts of electric generation capacity (utility-scale). Hydro is conventional hydroelectric. Petroleum includes petroleum liquids, petroleum coke, other gases, hydroelectric pumped storage, and other sources.

The U.S. currently relies on 92 nuclear plants, 1,900 natural gas plants, 240 coal plants, 80,000 wind turbines, and hundreds of square miles of solar panels for nearly 90% of its electricity. However, closures of nuclear and coal plants, coupled with increased reliance on weather-dependent wind and solar, is endangering grid reliability and affordability.

Electricity shortages and blackouts are predicted in two-thirds of the U.S., driven by the shutdown of stable, full-time power sources like coal. We require sufficient on-demand power. Without it, we will suffer.

Affordable, reliable electricity is no longer guaranteed, as climate-driven policies inflate energy costs and compromise grid resilience. The highest priced European countries have the greatest wind and solar penetration. Germany and Denmark pay nearly triple U.S. electricity prices. Californians pay double the national average, and it's growing fast. These are examples of how prioritizing climate ideology leads to ever-increasing prices and electricity insecurity. When you vote, you must look closely and prioritize electricity policies, prioritizing grid stability and affordability over climate agendas. Our economy and the affordability and reliability of our electricity depend on it.

ADDING WIND AND SOLAR MAKES YOUR ELECTRIC BILL GO UP, UP, UP!

Imagine you bought a solar car that worked like solar panels on the grid. You're excited because the salesman said you'd save a lot of money by not having to pay for gas. In addition, the government rebates about one-third of what you paid for it. However, you then discover that the car doesn't work for the first and last hour of the day because the sun isn't high enough to power it, or when it is misty, pouring rain, or cloudy. Therefore, you keep the gas vehicle.

You are now paying for both a reliable gas and an unreliable solar car. Your car costs have clearly gone up. Then you buy a wind automobile, which works similarly to a wind turbine. But only when the wind is blowing between 12 and 45 miles per hour. You discover this is only about one-third of the time. Your car costs go up even more.

This is what happens when we add wind and solar to our grid: it drives up costs. The more we add, the higher the costs rise.

On the real electric grid, things are even worse. We must maintain and pay for natural gas, coal, and nuclear power plants, too. The finance costs, labor, maintenance, and fuel will be there no matter how much wind and solar we add, if we want to keep our hospitals open and our electricity running all the time. Part-time, undependable wind and solar cannot keep the lights on all of the time, and that's unacceptable when it comes to the business of saving lives.

All of these redundant energy sources are expensive. They also do not reduce CO_2 emissions as advertised. Because solar and wind towers contain a lot of steel, rare metals, and cement, they produce a lot of pollutants when manufactured and only last 25 years or less before needing to be entirely replaced! Full-time power must be kept operating at all times, whether or not it is producing electricity, because they cannot be switched off and back on when the wind ceases or it rains.

In stop-and-go traffic, your car burns more gas. As we use our full-time electricity plants part-time, they produce more emissions and provide less electricity since they are cranked up and down more frequently to match customer demand, along with wind, and solar fluctuations. This incurs higher costs and actually increases CO_2 emissions. Ironically, these are the very things the cult of climate change say wind and solar will save us from!

The on-demand plants are being used part-time because of more wind and solar power. This means that their per-unit cost of electricity rises, forcing them to charge more. Colorado now faces blackouts because they closed reliable, low-cost coal plants.

Rates across the U.S. could rise up to five times what they are today because of the obsession with climate dogma and the bad energy policies it demands. We must put a stop to bad energy policy!

Worse yet is the fact that taxpayers are subsidizing the building of these unsustainable mechanisms. We then leverage private and electricity users' money to buy them from our rival, CO_2-spewing totalitarian Communist China. Eighty percent or more of the solar panels, inner workings of wind towers, EV, and grid batteries are all made in China using coal (they don't use clean coal technology like we do because it is expensive and it reduces electricity output by 20%), and involve either forced or slave labor. Then they dump the toxic chemicals and metals needed directly into lakes, rivers, and streams.

This pollution is obviously not good for the Earth. It sends our rival (and largest threat to freedom) more and more resources while increasing our dependence on them. Adding to our $34 trillion national debt to increase our electricity costs. Does this seem like it makes sense? Not to mention, it destabilizes and causes blackouts on our grid. This is terrible policy. This is all while the cult of climate change flies around the world on private jets, like John Kerry, who has said, "even if we get to net-zero (no CO_2 emissions), it will not stop climate change[175]." If it will not change the future climate or weather, why waste the money?

GREEN HYDROGEN: ANOTHER BOONDOGGLE OF WISHFUL THINKING

> *"Burning hydrogen is an economic and environmental mistake. Hydrogen is a luxury product and should not be used for combustion, no matter what politicians say... Burning hydrogen for energy purposes is like burning Louis Vuitton bags."*
> Samuel Furfari
> From his book *The Hydrogen Illusion*, pages 94–95
> https://www.amazon.com/hydrogen-illusion-Samuel-Furfari/dp/B08KHGDZNS

Professor Samuel Furfari is a chemical engineer from the Free University of Brussels. For 36 years, he was a senior official at the Energy Directorate General of the European Commission. He was Professor of Geopolitics and Energy Policy for 20 years. He is the author of 18 books on energy and sustainable development. He was President of the European Society of Engineers and Industrialists.

---

[175] https://nypost.com/2021/01/27/kerry-zero-emissions-wont-make-difference-in-climate-change/

Green hydrogen is a boondoggle because of the energy and water it takes to produce it. It takes thirteen times more water than hydrogen produced. Groundwater or river water must be purified, and ocean water must be desalinated. All of this adds to the cost. What's more, the water is then heated to 2,000 degrees and electrocuted. This yields hydrogen; and the entire process is quite mind boggling and, plainly, ridiculously expensive.

The new hydrogen must be frozen to near absolute zero, more than 450 degrees Fahrenheit lower than zero on a Fahrenheit scale. Then it must be compressed to ten thousand psi, which is about triple your standard scuba tank. Now you have liquid, super-chilled, and compressed hydrogen. Obviously, it must be stored somewhere safe until it is used, because it will explode if not handled properly.

Hydrogen cannot be piped easily because it is the world's smallest molecule[176] and escapes cracks very easily. It also embrittles any metal that it comes in contact with except for aluminum. This means that it will corrode and crack most metals and, when it escapes, if it doesn't float away, it is very likely to explode.

There are a lot of problems with hydrogen, not just the energy cost of making it. It is estimated that the energy to make hydrogen costs about 40% of the energy that is contained in the hydrogen made. Then, when it is used, another 20% is lost in its use. So, if this created hydrogen is used to make electricity to make more hydrogen, it will only net 20% of the original energy used to make it. This is a bad deal.

It clearly makes far more sense to use natural gas. Natural gas is inexpensive and clean. Of course, the cult of climate change doesn't like natural gas because it also turns into CO_2 when combusted for energy. We already have a way to move and use compressed natural gas (CNG). It works similarly to the way gas works in a car; the infrastructure is already created.

A common talking point is the concept that when you have excess wind and solar you can use it to create green hydrogen. This is simply magical thinking. The industrial processes required to create "green hydrogen" cannot be turned on when there's a lot of wind or sun and then turned off when it gets dark, or the wind stops. They must be running continuously.

---

[176] https://www.sciencedirect.com/topics/engineering/hydrogen-molecule

The climate propagandists tell us that they want to use hydrogen to solve the intermittency problem of wind and solar, but that is also a problem in creating hydrogen. The cult of climate change clearly doesn't care about the real-world costs; they have an agenda, as I laid out in Part One of this book. It's not about being practical; it's ultimately about total control. After all, green hydrogen costs about five times more than making it with natural gas. Somebody is profiting, and it's not the average non-billionaire like you and me.

There's no reason to make green hydrogen if we can use natural gas instead. Additionally, using hydrogen to generate electricity in order to keep hydrogen factories running when there isn't any wind or solar would be incredibly expensive. By driving up the cost of oil, coal, and natural gas exponentially, green energy "solutions" will be more competitive, at a dire cost to the majority of consumers in need.

This is what is meant by a circular economy: creating a shrinking circle. It implies using more and more expensive energy to create energy. The climate propagandists also talk about using hydrogen instead of natural gas to generate heat for thermal electricity. No one has done this at scale, and hydrogen will not work in the same way that natural gas does in creating the heat to spin the giant turbines. They are not interchangeable.

ELECTRIC CAR BATTERIES

An average electric car (EV) battery weighs around 1,000 pounds, made up of approximately 25 pounds of lithium, 65 pounds of nickel, 45 pounds of manganese, 30 pounds of cobalt, 200 pounds of copper, and 400 pounds of aluminum, steel, and plastic. Inside, there are about 7,000 lithium cells. The production of these materials requires about 500,000 pounds of rock to be unearthed and processed.

The Democratic Republic of Congo leads global cobalt production, accounting for roughly 70% of the world's total output, about 100,000 tons[177]. We will need ten to thirty times more cobalt for an all-electric car

---

[177] https://www.statista.com/statistics/339834/mine-production-of-cobalt-in-dr-congo/

future—every year. That's a lot of mining. Mining is often destructive, particularly in countries like the Congo where there are no environmental protections. And it is costly to restore the land after mining. If it is restored, lakes are often created. There are lakes that people have cottages on, that are clean and look good, that were former mines.

Unfortunately, at least 40% of this cobalt is mined under harsh conditions, with child laborers sometimes as young as six or seven. Most consumers overlook the origin and ethical concerns surrounding cobalt mining. Cobalt is blood metal, like diamonds are blood diamonds. Consumers should demand an ethical change to end the horrendous mining conditions and the use of children. Implementing environmental and labor regulations will increase production costs; it will add to the cost of batteries and EVs.

Communist China is the dominant producer and processor of cobalt needed for batteries. Apart from a single cobalt refinery in Finland, all other cobalt refining is controlled by China. Because of its large ownership percentage of the mines in Congo, it has considerable control over mining operations.

China refines cobalt and other necessary metals using coal electricity without the clean coal technology that we use. They use forced labor for some, and they have almost no environmental protections. This is why they are the low-cost provider. They have cornered the market. What happens if they cut off the supply to the rest of the world? What if they do this only once we become fully dependent on them, if and when we are only making EVs and are no longer manufacturing gas cars? After we have ripped up all the factories that make gas cars?

Dependence on China poses many risks, especially considering geopolitical tensions such as the potential conflict over Taiwan. With China being a major chip producer, any disruption could have severe consequences. The green energy movement, reliant on wind, solar, and batteries, depends on China for critical materials and manufacturing.

China has 75% of the world's largest battery factories, while the U.S. only has 5%. It's logical for China to prioritize its own battery production needs, limiting the availability of batteries for other markets. They are already doing this with the needed graphite, which is 100% processed in China. They have throttled back exports and may ban them altogether. It is unwise to give a totalitarian regime, which has been proven to use its power for their own gain and doesn't care about their trading partners, let alone their rival, that kind of power over our economy and wellbeing.

Hybrids make so much more sense. They can make 60–90 hybrid car batteries from just one EV battery. They also have a range similar to gas cars. They come in the plug-in and the non-plug-in variety. When the battery runs out, they run on gas generation. They work the way most trains work in the world. The gas or diesel engine makes electricity to run a more powerful and efficient electric motor.

Hybrids weigh a lot less, so our roads and parking garages will not suffer. And their tires will last longer, like gas cars. EV tires can wear out up to five times faster because of the weight. Studies show that tire wear is the leading cause of nano-plastics and particulates in the air in cities. The faster wearing EV tires will only add more of this pollution. If people can't afford them or don't buy them, we will not have this problem. The globalists want 100s of millions fewer cars on the road. Your car is in their sights.

https://www.weforum.org/agenda/2016/12/goodbye-car-ownership-hello-clean-air-this-is-the-futfure-of-transport/

Each electric car is subsidized with almost $50,000 of direct tax money and through regulations. And they still cost more! Without the subsidies, hardly any would have been sold. The way Tesla has been able to make billions is by selling their mph credits to other car makers. Each car maker has a fleetwide mph per vehicle standard. They average this across all the cars they sell. https://www.texaspolicy.com/wp-content/uploads/2023/10/2023-10-TrueCostofEVs-BennettIsaac.pdf

Ford wants to sell as many of the most popular vehicle in America—the Ford F-150 as they can and are allowed to under the mpg fleet standards. But they have to average their 12 to 23 miles per gallon (mpg) of each F-150 they sell with all the other vehicles they sell. They sell F-150s at a much higher price and profit margin than a Ford Echo. Because regulations will not let them sell as many F-150s as the world wants to buy. They must offset their below mpg F-150s and other low mileage cars with cars with mpg above the regulations required 40 miles mpg fleet standard.

They may even sell an Echo at a loss or no profit, just to balance the mpgs for the F-150s they are selling to meet the government fleet standard. But it doesn't get them enough MPGe credits. So, they buy them from electric vehicle makers, like Tesla, for billions.

The regulatory scam is even worse than this. EV makers get to say their EVs get 113 mpg. Cars just like so many other regulations are incredibly complex. They also have a complicated formula for increasing the mpg and investments. Which allows some EVs to have as much as 507 MPGe worth of credits. This is an incredible scam we are all paying for which makes our gas cars so much more expensive. If you are really interested in this check out the source I have provided. https://www.texaspolicy.com/wp-content/uploads/2023/10/2023-10-TrueCostofEVs-BennettIsaac.pdf

So, they get to sell MPGe credits worth 113 to 507 MPGe per vehicle. Without this behind-the-scenes gas powered cars provided subsidy by regulations, each EV would cost about $50,000 more. Few people could afford them. The EV companies wouldn't be in business. Besides this being hidden from the public and a direct transfer of money from gas car buyers to EV makers, it drives up the cost of every new car in America, except EVs. And because new cars cost more, used cars cost more. The average car on American roads is more than 12 years old and getting older. This is a scam, and it is wrong.

Think about this: what happens to the cost of F-150s, or any gas car, once the MPGe gets to an amount that can't be offset by buying EV MPGe credits? First, the cost of gas cars will skyrocket. Then after none of them are around to subsidize EVs, their prices will skyrocket too. This is the evil plan of the globalists and leftist politicians, like Joe Biden, to end private car ownership for most regular Americans. This will take some years and happen slowly. If we do not change directions at the ballot box, it will happen. There still is time.

On top of this, our electric grid is already short of electricity, and it is getting shorter all the time because Biden is attempting, through regulations, to close the coal plants that provide 20% of our electricity. And he is proposing to make natural gas plants use unproven expensive carbon capture (CCS) to capture CO_2 from them, and pipeline it somewhere, and pump it far underground…hoping it will never pop up again.

We can't build wind and solar fast enough and there is no affordable economic or engineering solution to their unreliable, intermittency problem. Biden is forcing this EV transition and there isn't enough electricity to charge all the additional EVs and all the AI data centers that need even more electricity than EVs. Guess where all the transformers needed to upgrade and maintain our grid with rising demand come from? I bet you guessed it: China. Yes, they've cornered this manufacturing market, too. Greater demand triggers upstream upgrades to support the increased electricity demand. Increasing the cost for all electricity users.

We need to change directions fast…Before it is too late!

ESG

Environmental, Social, and Governance (ESG) has wormed its way into most major corporations, with banks pushing these leftist non-financial standards on mid-size and small businesses. They impose leftist ideology onto businesses without elections or, in most cases, without our knowledge. Non-compliance gets you higher loan rates or no loans at all. You can even have your bank accounts terminated without warning; a fact that has prompted pushback from nearly twenty state Attorneys General.

ESG initiatives, including the pursuit of net-zero pledges, are costly and inefficient, diverting resources from shareholder interests and business growth. ESG is like the Chinese Communist Party's social credit system, except for businesses. Bend your business's knee to "woke" ideals and get rewarded; dissent and be penalized.

ESG undermines the traditional corporate responsibility of maximizing shareholder value and profitability, where the primary focus is on serving the customer and the shareholder. As ESG replaces the primary purpose of companies, shareholder interests suffer, and so do their retirement funds, like 401Ks, because money, time, and resources are diverted to the unnecessary, expensive, impossible deCO$_2$ing priorities, and other leftist priorities by ESG.

No one can serve two masters. The main priority is serving your customers, growing your company, and shareholder value…isn't it? ESG has hijacked this. It is likely illegal, and hard to prove. Some shareholders should sue them for violating their fiduciary duty.

CONCLUSION

Now that you've learned about the climate and energy lies at play and the variety of ways that they're being weaponized to wreak havoc on the world, I hope I've presented enough evidence to convince you that the concept of "man-made catastrophic climate change" is a tool designed for ultimate human despair. The cult of climate change is an undeniable pawn in the globalist's nefarious power play to control us all. Climate change regulations will do considerable damage to our global economies and, most importantly, our Earth.

We deserve to have low-cost, reliable energy and affordable, abundant food of our choosing. We should also be able to drive cars of our choice, fly on planes, and go where we want when we want, run profitable, sovereign businesses without political corruption impeding our success, and most importantly, to develop truly sustainable practices that don't farm out labor to corrupt totalitarian Communist China, at the tremendous harm of the environment.

Above all, rejecting the cult of climate change is a matter of national security. If we fail to elect representatives who will lead us in the direction of common sense—*not* these climate and energy lies—we will end up with globalist control, having their lies and nefarious agendas forced on us, whether we like it or not. We will see a social credit score system forced on us, and monthly CO_2 rations linked to digital currencies that can be turned off and controlled. They will also force us to keep our dissenting thoughts to ourselves or see our scores lowered and our ability to do anything shut down.

If you don't play their game, you'll pay with your freedom.

The social credit score, not the CO_2 rations, have already been implemented in totalitarian China. China doesn't seem to believe in the climate propaganda scam, other than to make and sell us the wind, solar, batteries, and soon EVs for the so-called "green" agenda. By their rules, as long as you are a good communist, behave yourself, and never express a dissident thought or speak ill of the party or leaders, you can do what they let you.

Jaywalk, read your Bible, or speak badly about the totalitarian communist party, and you will find yourself unable to move, buy anything, or even access the internet. We must fight this future loss of freedom. We are being propagandized; they say it's a movement for the common good. I hope, as the blinders come off, you can see that we have been rallied up in arms for a climate that *does not need saving*.

The Australia Commonwealth Bank (CBA) has already partnered with CoGo, a "carbon management solutions" company that sends all of its credit and debit card users a monthly report on their CO_2 emissions from purchases, along with ideas on how to reduce them[178]. If we don't fight back against the climate and energy lies, this is the fate we will suffer. Except the future CO_2 reductions will not be suggestions...they will be requirements.

Everything we eat, do, or consume emits CO_2, even breathing. That's why CO_2 is such an effective control tool for all of us. Despite increased CO_2 being so beneficial for the planet. As you already know, left is right and up is down in this upside-down world of the, corrupt globalist agenda.

Most confoundingly, in this upside-down world, it is the average citizen who suffers, not the wealthy individuals, politicians, and government bureaucrats who will create the regulate restrictions; they can purchase or be allowed additional CO_2 indulgences or receive exemptions.

We regular people will run out or feel the pinch every month. Imagine being prohibited from celebrating your daughter's wedding, a funeral, or a big family get together because you have run out of your monthly CO_2 rations and can't afford to buy more.

As you can read in a plethora of other titles, such as *The Creature from Jekyll Island*[179] by G. Edward Griffin or *The Great Reset*[180] by Glenn Beck, the elites have been masterminding their corrupt agenda for more than 100 years. They are patient and unwavering; they want to rule all of us for their benefit, not ours. They seek to force us to embrace their ideas of how we

---

[178] https://climatechangedispatch.com/aussie-bank-now-linking-customer-transactions-to-carbon-footprint/
[179] https://a.co/d/9txjmY2
[180] https://a.co/d/cRUCI2J

should live, what we should consume, and how we should limit our lifestyles. They say it's for our own good, but it isn't.

The global elite would have us revert to serfdom, which is one reason their strategy is to undo a century of the Industrial Revolution; so that we might return to the days when monarchies ruled, and the common people suffered. They want to revert to a time before hydrocarbons became widely available, providing prosperity for the many; a time when regular people had little freedom and 90 percent of the population lived in poverty.

Klaus Schwab, the World Economic Forum's (WEF) leader, famously said, "You will own nothing and be happy." He implies that we'll be renting everything. Our homes, and everything within them. The owners will be large global corporations. Corporations will dominate our lives. Although the 2016 article entitled "8 Predictions for the World in 2020" has been scrubbed from the internet, you can find a Wayback Machine snapshot in my footnotes[181]. Since you've read this book, you might find a few of the predictions increasingly haunting. They are:

1. All products will have become services. "I don't own anything. I don't own a car. I don't own a house. I don't own any appliances or any clothes," writes Danish MP Ida Auken. Shopping is a distant memory in the city of 2030, whose inhabitants have cracked clean energy and borrow what they need on demand. It sounds utopian, until she mentions that her every move is tracked and outside the city live swathes of discontents., the ultimate depiction of a society split in two.

2. There is a global price on carbon. China took the lead in 2017 with a market for trading the right to emit a ton of CO_2, setting the world on a path toward a single carbon price and a powerful incentive to ditch fossil fuels, predicts Jane Burston, Head of Climate and Environment at the UK's National Physical Laboratory. Europe, meanwhile, found itself at the center of the trade in cheap, efficient solar panels, as prices for renewables fell sharply.

---

[181] https://web.archive.org/web/20161125144156/https://www.weforum.org/agenda/2016/11/8-predictions-for-the-world-in-2030/

3. U.S. dominance is over. We have a handful of global powers. Nation states will have staged a comeback, writes Robert Muggah, Research Director at the Igarapé Institute. Instead of a single force, a handful of countries — the U.S., Russia, China, Germany, India, and Japan, chief among them — show semi-imperial tendencies. However, at the same time, the role of the state is threatened by trends including the rise of cities and globalist controls of the United Nations and World Economic Forum. and the spread of online identities[182].

Because I'm fearful that we will even lose this Wayback Machine access to such doom and gloom predictions, I've screen captured the entire article below. This underscores the importance of purchasing physical books, even if you are currently consuming mostly audiobooks.

---

[182] https://web.archive.org/web/20220604040742/https://www.weforum.org/agenda/2016/11/8-predictions-for-the-world-in-2030/

As Brexit and Donald Trump's victory show, predicting even the immediate future is no easy feat. When it comes to what our world will look like in the medium-term – how we will organise our cities, where we will get our power from, what we will eat, what it will mean to be a refugee – it gets even trickier. But imagining the societies of tomorrow can give us a fresh perspective on the challenges and opportunities of today.

We asked experts from our Global Future Councils for their take on the world in 2030, and these are the results, from the death of shopping to the resurgence of the nation state.

1. All products will have become services. "I don't own anything. I don't own a car. I don't own a house. I don't own any appliances or any clothes," writes Danish MP Ida Auken. Shopping is a distant memory in the city of 2030, whose inhabitants have cracked clean energy and borrow what they need on demand. It sounds utopian, until she mentions that her every move is tracked and outside the city live swathes of discontents, the ultimate depiction of a society split in two.

2. There is a global price on carbon. China took the lead in 2017 with a market for trading the right to emit a tonne of CO2, setting the world on a path towards a single carbon price and a powerful incentive to ditch fossil fuels, predicts Jane Burston, Head of Climate and Environment at the UK's National Physical Laboratory. Europe, meanwhile, found itself at the centre of the trade in cheap, efficient solar panels, as prices for renewables fell sharply.

3. US dominance is over. We have a handful of global powers. Nation states will have staged a comeback, writes Robert Muggah, Research Director at the Igarapé Institute. Instead of a single force, a handful of countries – the U.S., Russia, China, Germany, India and Japan chief among them – show semi-imperial tendencies. However, at the same time, the role of the state is threatened by trends including the rise of cities and the spread of online identities,

4. Farewell hospital, hello home-spital. Technology will have further disrupted disease, writes Melanie Walker, a medical doctor and World Bank advisor. The hospital as we know it will be on its way out, with fewer accidents thanks to self-driving cars and great strides in preventive and personalised medicine. Scalpels and organ donors are out, tiny robotic tubes and bio-printed organs are in.

5. We are eating much less meat. Rather like our grandparents, we will treat meat as a treat rather than a staple, writes Tim Benton, Professor of Population Ecology at the University of Leeds, UK. It won't be big agriculture or little artisan producers that win, but rather a combination of the two, with convenience food redesigned to be healthier and less harmful to the environment.

6. Today's Syrian refugees, 2030's CEOs. Highly educated Syrian refugees will have come of age by 2030, making the case for the economic integration of those who have been forced to flee conflict. The world needs to be better prepared for populations on the move, writes Lorna Solis, Founder and CEO of the NGO Blue Rose Compass, as climate change will have displaced 1 billion people.

7. The values that built the West will have been tested to breaking point. We forget the checks and balances that bolster our democracies at our peril, writes Kenneth Roth, Executive Director of Human Rights Watch.

8. "By the 2030s, we'll be ready to move humans toward the Red Planet." What's more, once we get there, we'll probably discover evidence of alien life, writes Ellen Stofan, Chief Scientist at NASA. Big science will help us to answer big questions about life on earth, as well as opening up practical applications for space technology.

Images Courtesy of the Wayback Machine Archive[183]

To summarize the terrifying predictions above: we shall be rewarded as long as we obey and continue to be good citizens as defined by the World Economic Forum. American dominance and sovereignty shall be squelched. We don't own anything, and we'll be charged for CO_2 usage. Western values will be eliminated.

This is the series of predictions as told straight from the World Economic Forum website, an unelected organization that has captured the compliance of most major economic movers and companies. For 50 years, the global elites have gone to Davos, Switzerland, annually. Often arriving by private jet and taking the last leg by helicopter. They are too important to be concerned with CO_2 emissions, which is for the peons they conspire to control.

Rather than targeting communist China for their suppression of human rights, human slave labor, prison camps, and environmental pollution, the WEF ideals coincide with and even resemble those of the Chinese totalitarian regime. Worse yet, the WEF has global adherence and compliance among policymakers. Which means that the totalitarian ideologies are making their way into your community. Many of them already have.

Before you can say, "No way! You're making this up. It couldn't happen here," then consider this one story of a man named Nigel Farage, the well-known Brexit campaigner. He led the campaign for Britain to leave the E.U. That was a big setback for the globalists and their goals. First, Nigel was de-banked[184]; his bank of 40 years informed him that his accounts would be canceled due to his public opposition to net-zero, the E.U., and other beliefs those financial institutions didn't support. Farage thought, "I'll get a new bank."

---

[183] https://web.archive.org/web/20220604040742/https://www.weforum.org/agenda/2016/11/8-predictions-for-the-world-in-2030/
[184] https://www.reuters.com/world/uk/farage-makes-fresh-allegations-against-uks-coutts-over-account-closures-2023-07-19/

But Farage was summarily rejected by ten other banks. He was then able to obtain the internal records of the conspiring banks that led to his de-banking. They claimed, in short, that Farage was against their ideals, and was too outspoken. Presumably, they also assumed he'd be too embarrassed to go public and if he did it wouldn't make any difference. Boy, were they wrong...

Farage went very public. He was able to get two bank CEOs fired, and the largest bank in England saw its stock value plummet by 30% once he made the de-banking coup public. He was able to locate a bank that would let him open an account. He had the fame and savvy to fight back. Regular folks like us do not have the ability to fight back the way Farage did. If they did it to Nigel, they can do it to us, and be much more successful.

What if you couldn't get a bank account, a mortgage, or a credit card?

The only way to achieve net-zero, or no CO_2 emissions, is to micromanage, and potentially de-platform or de-bank anyone they want to. Adherence to the globalist agenda means that each of us must consume significantly less or nothing. That we obey. That we shut up with any questions or opposition of their narrative. That we bend the knee in all things. That we accept their control or else.

Most frighteningly, these globalists are parrots for the "depopulation" agenda. They believe that 8 billion people are too many for Earth. I strongly disagree.

How are "they" going to choose who of "us" to eliminate?

Just starve some with food shortages? Remember, each of us exhales about 700 pounds of CO_2 a year, which, combined with praising "depopulation" and "overpopulation" rhetoric, has sparked many a meme with the theme, "you are the carbon they want to reduce."

The WEF-affiliated C40 cities[185] have agendas which require every one of their citizens to do the following. You should check to see if your city is a member. Ask them why. Expose them. See if they give you the cover up response that *USA Today* did, after this came to light, claiming that they aren't goals, just aspirations.

1. One may only buy three sets of clothes every year

---

[185] https://www.c40.org/cities/

2. There shall be no more automobiles for average people (everything you need shall be within a "15-minute" walking or biking distance, hence the moniker "15-minute cities"). Oxford England is attempting to implement 15-minute cities now.
3. Plane trips shall only be allowed once every four years, if at all.
4. Citizens shall go vegan or consume as little meat as possible. They are also advocating the idea of eating bugs instead of meat for protein.

In summation, please use this book as a guide to deprogramming yourself and anyone else who will listen to you. Use it to help others who have not joined the climate cult understand the enormity of the problem we face. Help them understand the facts. If we do not, all will be lost.

The moment to act is now! Before it's too late, we must be resolute to help our brothers and sisters wake up. Help them wake up from their propaganda, indoctrination, and programmed climate and energy delusion. I believe we should cease voting for the primarily left-leaning politicians who support the cult of climate change's agenda and lies. They are all the Democrats and some Republicans. You must use discernment. If they are labeled "climate deniers" by the propaganda media, it is a good bet they are on our side.

The climate cult "solutions" for a problem they have made up to frighten us, fail to stand up in the real world, they're unnecessary and, in most cases, much worse for our Earth. They will drive our energy costs through the roof, cause everything else to become more expensive because energy is in all that we do or consume, and threaten our national security, our freedom, and our very way of life.

Let's activate the spreading of this knowledge and truth, before it's too late.

INDEX

Symbols
9/11 47

A
Alexander, Dr. William 231
alteration, of data 87
Ardis, Dr. Brian 46
Argentina 93, 135
Aristotle 49
Australia 63, 95, 121, 170, 176, 178, 236, 255, 299

B
Ballantyne, Ashley 84
Balling Jr., Robert C. 165
Bastardi, Joe 215
beef 31
Biden, Joe 46, 238, 239, 241, 296, 297
Brown, Patrick T. 236, 237
Bryson, Dr. Reid 149, 150, 151

C
carbon-neutral 22
censorship 44
central planning 10, 17, 18, 47, 48
China 9, 16, 30, 32, 47, 48, 50, 131, 256, 270, 272, 274, 275, 276, 277, 278, 281, 283, 290, 294, 297, 298, 300, 301, 303
Clauser, John Francis 21
Climate Change Industrial Complex 7
ClimateGate 129, 130
Clough, Dr. Charles 195
CO2 Coalition 35, 37, 40, 59, 76, 187
coal 9, 16, 21, 22, 31, 32, 56, 84, 131, 249, 256, 261, 267, 270, 274, 275, 276, 277, 279, 280, 281, 282, 283, 284, 285, 286, 287, 288, 289, 290, 293, 294, 296
cocaine 51
coffee 31, 72
cold (illness) 248
cold (temperature) 55, 61, 62, 101, 103, 108, 109, 154, 159, 161, 162, 173, 190, 192, 205, 212, 213, 216, 247, 248, 249, 277, 287
Congo viii, 141, 143, 144, 145, 255, 293, 294, 307
Cook Paper 42
Critical Race Theory (CRT) 48
Crockford, Susan 200, 202, 205
Cuba 47, 50, 278
cult 13, 18, 21, 23, 24, 27, 28, 29, 30, 31, 32, 34, 40, 41, 62, 68, 76, 78, 79, 83,

85, 122, 130, 147, 199, 203, 263, 277, 280, 287, 290, 291, 292, 293, 298, 305
Cunningham, Walter 33, 34, 110

D
decarbonization 18, 55
deforestation 257, 258
Dominican Republic 10, 257, 258
Dowdle, Dr. Watler R. 252
drought 77, 155, 228, 229, 230, 251
Duvat-Magnan, Virginie 180
Dyson, Dr. Freeman John 55

E
Economic Freedom Index 18
electricity 6, 9, 16, 249, 256, 261, 267, 271, 277, 279, 280, 281, 282, 283, 286, 287, 288, 289, 290, 291, 292, 293, 294, 295, 296, 297
El Niño 116, 117, 162
ENSO 98
Epstein, Alex 259

F
Farage, Nigel 303, 304
fertilizer 75, 76, 199, 262, 263, 287
food shortage 2, 77, 259, 304
food supply 6, 11, 16, 18, 22, 24, 31, 77, 167, 263, 277
forest 8, 31, 56, 122, 192, 193, 237, 238, 239, 242, 243, 255, 257, 258
free-market capitalism 17

G
Galilei, Galileo 49

gaslighting 7, 23, 78, 79, 183
glacier 172, 189, 190, 191, 192, 193, 195, 198, 199, 213
global cooling 119, 127, 130, 132, 161
globalist 2, 7, 11, 23, 57, 79, 263, 295, 296, 298, 299, 301, 303, 304
global warming 3, 25, 34, 40, 42, 57, 58, 62, 86, 88, 91, 92, 93, 94, 95, 97, 100, 117, 119, 124, 126, 127, 129, 131, 132, 146, 161, 165, 166, 167, 177, 182, 190, 201, 209, 210, 213, 231, 247
Google 186, 210, 211
Great Reset, the 6, 299
Greenhouse Effect, the 57, 59
Greenland 101, 108, 173, 190, 192, 195, 196, 197, 198, 199, 200
Guterres, António 168

H
Haiti 10, 257, 258
Hansen, James 125, 155, 156, 169, 182, 209, 210
Happer, Dr. William 57, 58, 63, 64, 65, 99
Heller, Tony 124
Heritage Foundation 18
hurricanes 31, 33, 91, 217, 218, 219, 220, 221, 222, 223, 224, 249, 250

I
Ice Age 2, 8, 27, 104, 149, 165, 191, 262
ice caps 101, 167, 172, 199
Idso, Dr. Craig 69, 70, 74
Idso, Dr. Sherwood 74
Intergovernmental Panel on Climate Change (IPCC) 33, 39, 56, 76, 85,

86, 87, 88, 89, 91, 95, 165, 180, 189, 190, 198, 206, 207, 227, 231, 250
Itoh, Dr. Kiminori 86, 87

J
Jones, Phil 129, 130

K
Kaser, Georg 189
Kiribati 181
Koonin, Steven 218

L
La Niña 162
Lasee, Frank v, 61
Legates, David 42, 43
Lindzen, Dr. Richard 18, 23, 44, 58, 64
Lomborg, Bjorn 247, 249, 251
Lysenko, Trofim 48, 50

M
Mann, Michael 95, 130
manufacturing 283, 287, 294, 297
Martz, Chris 45
McCullough, Dr. Peter 46
mind control 24, 47, 89
Moniz, António Egas 51

N
Nakamura, Dr. Mototaka 97, 100
natural gas 9, 16, 21, 22, 31, 32, 56, 61, 74, 76, 84, 249, 256, 262, 267, 270, 274, 275, 279, 280, 283, 284, 285, 286, 287, 288, 289, 290, 292, 293, 296
new world order 169

Northrup, Dr. Christianne 46
Nova, Joanne 121, 122, 126, 127
nuclear power 16, 288, 290

O
Obama, Barack 42, 89, 169, 218
oil 9, 21, 22, 31, 32, 40, 56, 84, 218, 249, 267, 270, 274, 275, 279, 280, 281, 283, 284, 285, 286, 287, 293
Orwell, George 39

P
Pasteur, Louis 252
Pielke Jr., Dr. Roger 171, 224
planet 17, 30, 44, 69, 71, 91, 94, 97, 100, 108, 159, 168, 176, 184, 190, 192, 206, 213, 215, 231, 247, 286, 299
polar vortex 161, 162
Pol Pot 47, 278
Polvani, Lorenzo M. 211
PragerU 35, 36
propaganda v, 2, 7, 8, 11, 15, 22, 24, 26, 28, 35, 44, 46, 51, 55, 57, 62, 68, 72, 77, 100, 101, 103, 127, 149, 156, 161, 162, 168, 175, 179, 181, 203, 211, 236, 237, 238, 241, 253, 254, 272, 298, 305

R
Rankin, John E. 169
Reef, Coral 175, 176, 177, 178, 179, 180, 182
Reef, Great Barrier 175, 176, 177, 180
rice 31, 71, 72, 73, 75
Rice 73

Roser, Max 272
Russia 48, 50, 256, 286, 301

S

Santa Cruz Sentinel 128, 129
Schmitt, Harrison Hagan 67, 227, 228
sea levels 101, 165, 166, 168, 169, 170, 171, 172, 173, 181, 182, 183, 195, 197, 198, 208, 210, 213
Senate 8, 89, 91, 94, 104, 156
Shellenberger, Michael 15
Singh, Hansi A. 211
Smil, Vaclav 269, 270
socialist 18, 50, 168, 277, 278
solar 9, 10, 16, 21, 22, 31, 40, 49, 55, 72, 77, 86, 91, 96, 122, 149, 213, 215, 231, 261, 267, 269, 270, 271, 272, 274, 275, 276, 279, 287, 288, 289, 290, 292, 293, 294, 297, 298, 300
South Wales 134
Spencer, Dr. Roy W. 39, 98, 99, 116
Starck, Walter 175
Statue of Liberty 169
sun 49, 60, 63, 93, 94, 97, 98, 101, 108, 109, 203, 215, 216, 217, 231, 289, 292

T

Tans, Pieter 84
Tenpenny, Dr. Sherri 46
theory of relativity 48, 90
tornadoes 217, 224, 225, 226, 227, 250
totalitarian 9, 16, 23, 30, 48, 50, 290, 295, 298, 299, 303
Tuvalu 181, 182

U

United Nations (UN) 11, 24, 32, 39, 40, 57, 72, 76, 78, 84, 86, 91, 95, 107, 166, 167, 168, 169, 178, 181, 182, 183, 189, 211, 231, 250, 258, 262, 263, 301
Uruguay 136

V

Vanuatu 181
Venezuela 47, 50, 138, 278

W

Wadhams, Peter 208, 209
Washington Post, the 240
Wigley, Tom 129
wind towers 9, 10, 16, 122, 288, 290

Y

YouTube 63, 209, 211

Z

Zelenko, Dr. Vladimir 252, 253

PIERUCCI PUBLISHING

ELEVATING WORLD CONSCIOUSNESS THROUGH BOOKS.

More Info

www.PierucciPublishing.com

855-720-1111

Special Report

EMERGENCY MEDICAL KITS NOW AVAILABLE

TAKE ADVANTAGE OF SPECIAL READER DISCOUNTS WITH COUPON CODE "STEPHANIE"

"I don't know what would have happened if I didn't have the IVM in my cabinet when I got sick... and to think my local doctor wouldn't even consider a prescription for Ivermectin! But when I fell ill due to a crippling stomach bug, I knew that if I wanted to make it through another day, I needed help. Thankfully...

LEADING CARDIOLOGIST APPROVES OF NEW ER MED KITS

GOVERNMENT THREATENING TO BLOCK SALES OF LIFE-SAVING TREATMENTS...

The best way to avoid the hospital during an emergency is to have life-saving medications on hand. That's why The Wellness Company is now providing Emergency Medical Kits so that you and your family will have 8 Safe & Effective Emergency Medications on hand for a variety of health concerns. But the government doesn't want the public to have these 8...

MORE INFORMATION AT WWW.TWC.HEALTH/STEPHANIE

www.ingramcontent.com/pod-product-compliance
Ingram Content Group UK Ltd.
Pitfield, Milton Keynes, MK11 3LW, UK
UKHW021252180426
11946UKWH00004B/89